International Conference on Ship and Offshore Technology (ICSOT Indonesia 2012)

Developments in Ship Design & Construction

Ambon, Indonesia
7 – 8 November 2012

ISBN: 978-1-62993-038-1

Printed from e-media with permission by:

Curran Associates, Inc.
57 Morehouse Lane
Red Hook, NY 12571

Some format issues inherent in the e-media version may also appear in this print version.

Printed by Curran Associates, Inc. (2013)

For permission requests, please contact The Royal Institution of Naval Architects]
at the address below.

The Royal Institution of Naval Architects
10 Upper Belgrave Street
London SW1X 8BQ UK

Phone: 44 207 235 4622
Fax: 44 207 259 5912

publications@rina.org.uk

Additional copies of this publication are available from:

Curran Associates, Inc.
57 Morehouse Lane
Red Hook, NY 12571 USA
Phone: 845-758-0400
Fax: 845-758-2634
Email: curran@proceedings.com
Web: www.proceedings.com

CONTENTS

NUMERICAL SIMULATION INTO THE DRAG CHARACTERISTICS OF SYMMETRICAL AND ASYMMETRICAL CATAMARANS WITH VARIOUS DEMIHULL SEPARATIONS

A. Jamaluddin, Indonesian Hydrodynamic Laboratory (IHL) Surabaya – Indonesia
I K A P Utama, Dept. of Naval Architecture and Shipbuilding Engineering, ITS, Surabaya – Indonesia

SUMMARY

This paper describes the numerical prediction of resistance for symmetrical and asymmetrical catamaran with various demihull separations (S/L). Numerical simulations have been carried out to determine the incompressible free surface flow around the demihull form for which experimental results are available. The model involves the numerical solution of the Reynolds Averaged Navier-Stokes (RANS) equations along with the Volume of Fluid (VOF) method for predicting the free surface. The shear stress transport (SST) turbulence model has been used in the viscous flow code and a tetrahedral unstructured grid was applied for meshing the computational domain.

The verification was based on the examination of the flow around the hull for range of speeds and by comparison of the results for resistance obtained by CFD simulations and by experiments and other published information. This study shows that numerical investigation appears to provide very good agreement as has been observed in the resistance curves showing that CFD methods may adequately deal with the problem of resistance interference produced by both demihulls.

Key words: CFD modeling, resistance, symmetrical catamaran, asymmetrical catamaran, hull separation

NOMENCLATURE

Demihull	One of the hulls which make up the catamaran
L, L_{WL}	Length on waterline (m)
b	Breadth of demihull (m)
T	Draught (m)
V	Volume (m^3)
Δ	Displacement (kg)
S/L	Lateral separation (clearance) ratio, between demihull centrelines
Fr	Froude number
Cb	Block coefficient
C_T	Total resistance coefficient
ρ	Density of water (kg m^{-3})
μ	Dynamics viscosity (kg/m/s)
CFD	Computational Fluid Dynamics
SST	Shear Stress Transport
WSA	Wetted surface area (m^2)

1. INTRODUCTION

Predicting the ship resistance of the configurations of catamaran forms during the initial designing stage is essential to develop a speed-fuel efficient hull form. This has been one of the motivations of attempting to estimate and predict the flow characteristics around ship hull (for this specific case: resistance) using CFD tools. Some works have been carried out for maritime application such as reported by [1, 2, 3, 4, 5, and 6]. However, there are still uncertainties to compute the viscous and wave resistances as individual component. The reason for this is attributed to the unsuccessful of free surface modelling during those periods. The free surface algorithm is presently available and occupied by recent CFD codes. Thus, the current work is focused on the development of free surface modelling with consideration to breakdown ship resistance components and in particular case, the catamaran resistance.

In the present study, the commercial CFD code used is ANSYS-CFX 11.0 which model involves the numerical solution of the Reynolds Averaged Navier-Stokes (RANS) equations along with the Volume of Fluid (VOF) method for simulating the free surface flow. The shear stress transport (SST) turbulence model has been used in the viscous flow code for meshing the computational domain. Then, the verification was based on the examination of the flow around the hull for range of speeds and by comparison of the results for resistance obtained by CFD simulations and by experiments and other published information.

The demihull (monohull) and catamaran were used for computational resistance prediction as a benchmark to gain understanding of free surface simulations in the range of Froude numbers of 0.2 – 0.65 with three separation hull distances (clearances), including the single demihull form.

2. MODELLING AND NUMERICAL SIMULATIONS

To solve the governing fluid equation, the fluid domain is divided into a finite number of cells and these equations are changed into algebraic form via a discretization process in which the finite volume method is used [7]. For the initial simulation the SST turbulence model was applied [8, 9].

The boundary conditions were imposed in the computational as follows: a velocity inlet boundary condition was used downstream. The flow velocity (equal to the model velocity) and the free surface elevation were fixed at the inlet. A pressure outlet boundary condition was used downstream. The hydrostatic pressure at the outlet was calculated assuming an undisturbed free surface. Smooth walls with a free-slip condition were assumed for the floor and the side wall.

For the two phase flow solutions, a pressure opening boundary condition was used for the top surface. It was assumed that no water reached the top surface. The opening was also used as the reference point for the pressure. The water was modelled as fresh water at temperature of 26°C (density= 999 kg/m^3, dynamic viscosity= 1.137x10^{-3} kg/m/s). The air was modelled with a molecular mass of 28.96 kg/kmole and a dynamic viscosity of 1.8x10^{-5} kg/m/s.

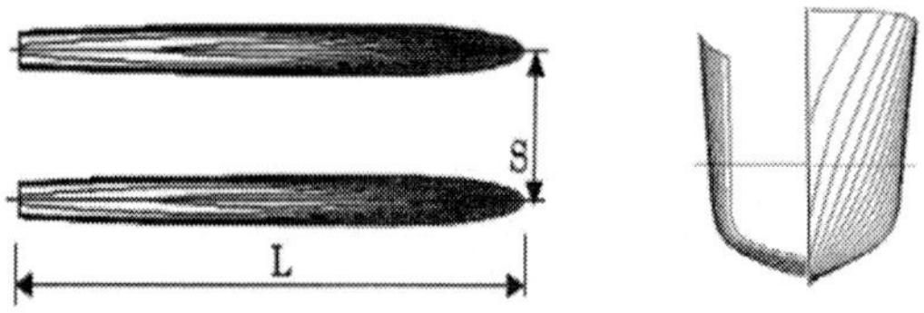

Figure 1. Symmetrical Catamaran Hull Configuration

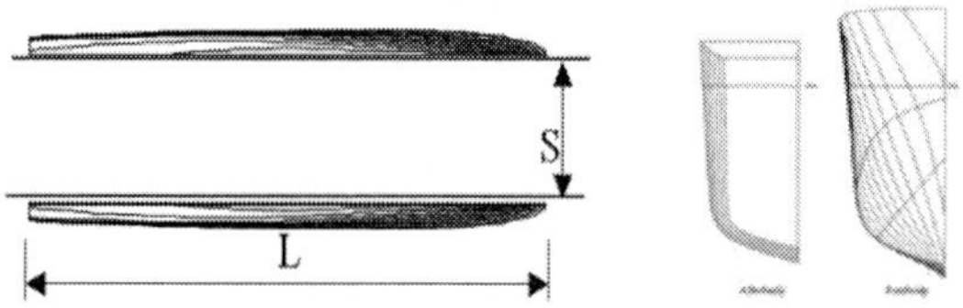

Figure 2. Asymmetrical Catamaran Hull Configuration

The symmetrical and asymmetrical catamaran hull forms were used for CFD simulations in the range of Froude numbers of 0.2 – 0.65 as shown in Figures 1 and 2. The main particulars of both models are presented in Table 1.

Table 1. Main Dimension of Demihull and Catamaran

Parameter	Demihull		Catamaran		Unit
	Asym.	Sym.	Asym.	Sym.	
L_{WL}	1.405	1.372	1.405	1.372	m
b	0.066	0.132	-	-	m
T	0.126	0.078	0.126	0.078	m
WSA	0.3563	0.2559	0.7127	0.5206	m^2
Volume	0.0070	0.0070	0.0140	0.0140	m^3
Displac.	7.0233	7.0233	14.0439	14.0439	kg

Figure 3 shows the initial computational domain. The mesh extends 1.5 model lengths upstream, 4 model lengths downstream, 1.5 model lengths sideways (for demihull and catamaran respectively) and sufficiently far above and below the free surface. The initial computational meshes used for multiphase flow calculations comprised with 822,087 and 1,304,943 mesh elements for demihull and catamaran respectively.

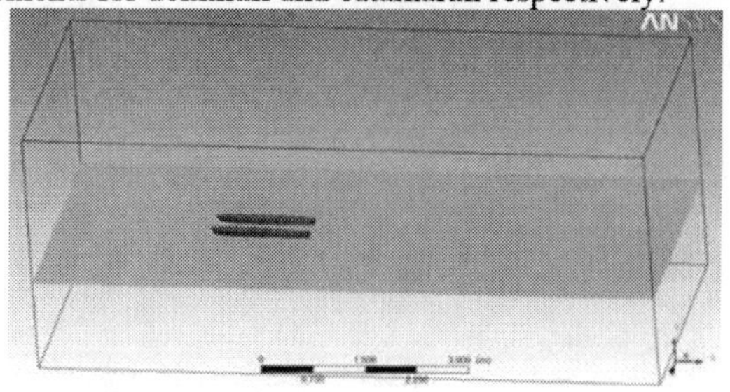

Figure 3. Initial Computational Domain

A tetrahedral unstructured grid was used with the viscous flow code for meshing the computational domain, while quadrilateral structural patches were used with the potential flow code for meshing the catamaran hull surface and the water surface around it. The grid quality is fundamental for the convergence and accuracy of CFD calculations. Grid qualities are discussed in detail by [10, 11]. The root-mean square (RMS) criterion for free surface simulation with residual target value (variable value) of 10^{-5} was used for checking the convergence of the solution. This target value is good convergence, and usually sufficient for most engineering applications [12].

3. COMPUTATIONAL RESULTS AND DISCUSSION

The CFD simulations were executed under the same conditions of the model scale tests to predict the resistances. This has the advantage that model scale results are available so that a direct comparison could be made without the need to consider scale effects or scaling techniques.

The total resistance of catamaran, which is composed of two bodies and placed in close proximity, is generally different from the sum of the resistance of the two bodies when tested separately. This phenomenon is called interference, and it arises from the change in flow pattern that accompanies the placing of the two bodies in close proximity [13]. The interference may be unfavourable or favourable.

The total resistance coefficients (C_T) are strongly affected by the change of hull separation (S/L) for both symmetrical and asymmetrical catamarans as shown in Figures 4 and 5. The interaction effects due to separation on the symmetrical catamaran are quite stronger, Figure 4, while these effects are less strong on the asymmetrical catamaran, Figure 5.

In general, it is shown that the smaller the clearance (S/L), the higher the resistances, particularly for symmetrical hull. This is attributed to the more intensive wave and viscous interaction between the hulls [14, 15]. For the maximum hull clearance (S/L= 0.4) is practically equal and slighly lower to double the demihull resistance at Fr <0.45, indicating that for this separation distance, the interaction effect has already vanished [16]. On the other hand, for Fr> 0.5, the resistance of catamaran hulls is lower than that of the double of the demihull for the given hull clearances. The reason for this is that the wave interference contributed a favourable effect [17, 18].

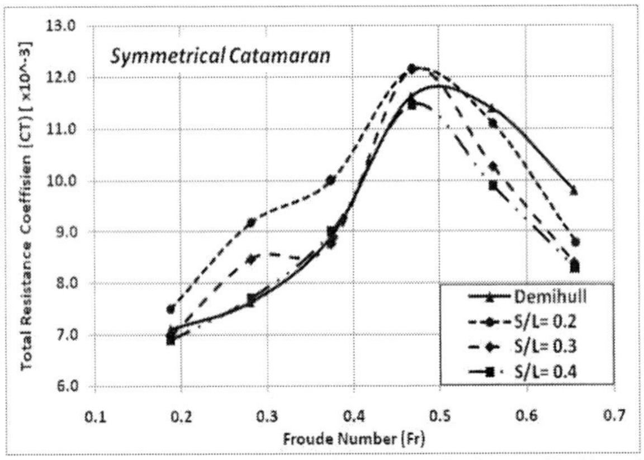

Figure 4. Total Resistances of Symmetrical Demihull and Catamaran

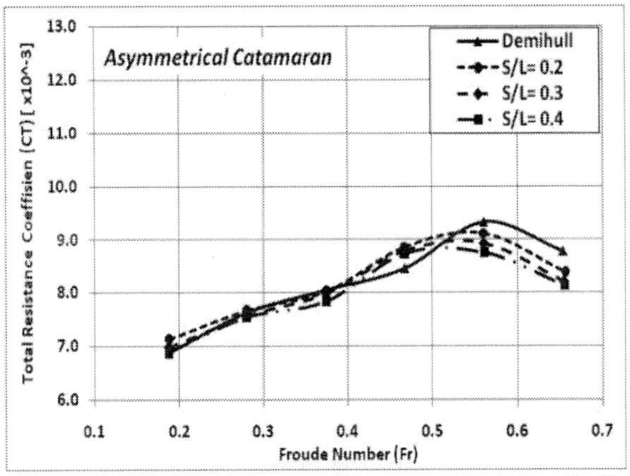

Figure 5. Total Resistances of Asymmetrical Demihull and Catamaran

The verification was based on the examination by comparison of the results for resistance obtained CFD simulations and by experiments in catamaran configuration.

Comparison of the computational and experimental data is shown in Figures 6 - 8. It can be seen that the agreement between computed values of the total resistances and the values provided by experiments is very good.

CFD simulation provides accurate results for resistance coefficients in the range of Froude numbers studied; this was also provided small percentage error between 0.5% - 6.0% or average error within 3% of the experimental data. Despite the differences of both results, the CFD proves a powerful design tool for free surface ship modelling to predict the ship resistance components and assessment prior to the final design validation model tests.

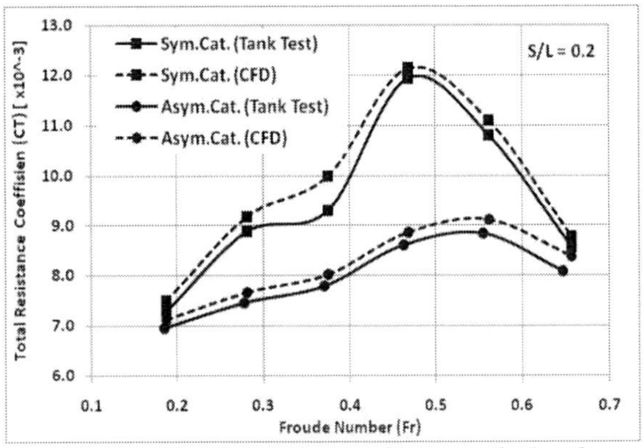

Figure 6. Total Resistances of Symmetrical and Asymmetrical Catamaran, S/L=0.2

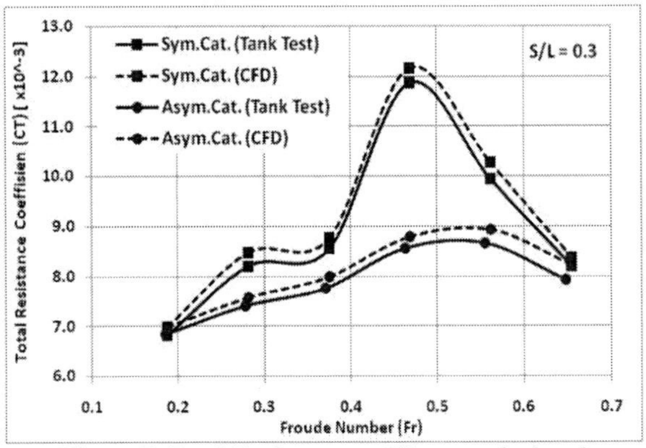

Figure 7. Total Resistances of Symmetrical and Asymmetrical Catamaran, S/L= 0.3

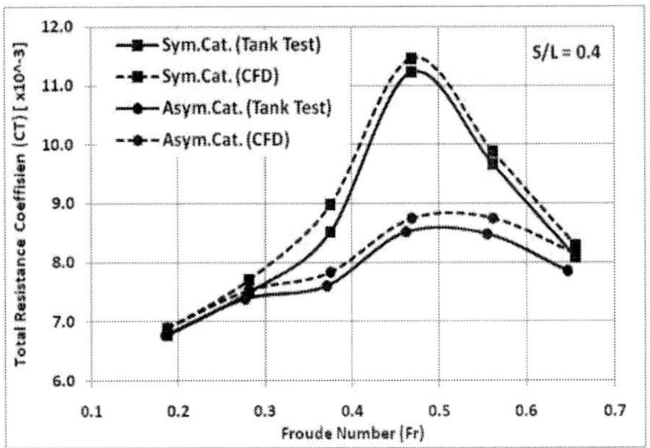

Figure 8. Total Resistances of Symmetrical and Asymmetrical Catamaran, S/L=0.4

The difference between numerical and experimental results may due to the errors of modelling, discretization and computation. Most of the time, computation error is a small part in the numerical error. When a discretization scheme is applied, discretization error is mainly determined by grid generation [11]. ITTC gave a simple suggestion on the research of grid uncertainty, but the suggestion still under discussion. The research about the uncertainty of the grid is an important part of numerical simulation, and has a direct relationship with the precision of the result.

4. CONCLUSIONS

The computational simulations are used to examine the resistance characteristics of symmetrical and asymmetrical catamarans. The computational predictions reveal similar tendencies with the experimental data, save for minor or small differences. The following conclusions can be drawn based on computations and compared to experimental data:

- The CFD analysis is shown that the hull clearance (S/L) for catamarans is strongly affected to the resistance characteristics. The general trend in all cases is that as the hull separation increases, the resistance decreases.
- The resistance are significantly affected on the symmetrical hull compared to the asymmetrical one. This is due to the wave and viscous interaction between the asymmetrical hulls is smaller.
- The CFD results have been shown to be significantly accurate, proving results for the resistance that are a few percent of the measured results. The stage has been reached where CFD analysis can stand on its own while evaluating various design configurations.

5. REFERENCES

1. Couser, P.R., Molland, A.F., Armstrong, N.A. and Utama, I.K.A.P. 'Calm water powering predictions for high-speed catamarans', *Proceedings FAST 97 Conference*, Sydney (Australia), July 1997.
2. Date, J.C. and Turnock, S.R., 'A study into the techniques needed to accurately predict skin friction using RANS solvers with validation against froude's historical flat plate experimental data', *Ship Science Report*, No. 114, Department of Ship Science, the University of Southampton, UK, March 1999.
3. Molland, A.F. and Utama, I.K.A.P., 'Experimental and numerical investigations into the drag characteristics of a pair of ellipsoids in close proximity, Journal of the Engineering for the Maritime Environment', *Part of the Proceedings of the Institution of Mechanical Engineers*, Vol. 216, No. M2, 2002.
4. Sahoo, P.K., Doctor, L.J. and Pretlove, L., 'CFD prediction of the wave resistance of a catamaran with staggered demihulls', *International Conference on Marine Hydrodynamics* (MAHI 2006), India, 2006.
5. Utama, I.K.A.P., Investigation of the viscous resistance components of catamaran forms, PhD Dissertation, Department of Ship Science, the University of Southampton, UK., 1999.
6. Utama, I.K.A.P., Setyawan, D., Murdijanto and Jamaluddin, A., 'Experimental and CFD investigation into the drag characteristics of catamaran fishing vessel', *Regional Conference on Mechanical and Aerospace Technology* (RCMAeT), Bali, Feb. 2010.
7. CFX, CFX Manual VII, Ansys 2007.
8. Menter, F.R., 'Zonal Two Equation k- ω turbulence Models for Flows', *AIAA Paper* 93 – 2906, 1993.
9. Menter, F.R., 'Two-Equation Eddy-Viscosity Turbulence Models for Engineering Applications', *AIAA Journal*, Vol. 32 (1994) 289-289.
10. Thompson, J.F., Soni, B.K., Weatherill, N.P., Handbook of Grid Generation, CRC Press. 1999.
11. Deng, R. Huang, D., Li, J., Cheng, X., Yu, L., 'Discussion of grid generation for catamaran resistance calculation', *Journal of Marine Sci. Appl.* (2010) 9: 187-191.
12. Dinham, T.A. Craddock, C. Lebas, A. and Ganguly, A., 'Use of CFD for hull form and appendage design assessment on an offshore patrol vessel and identification of a wake focussing effect', *RINA Marine CFD Confeference*, Southampton, UK., 2008.
13. Insel, M. and Molland, A.F., 'An investigation into the resistance components of high-speed displacement catamarans', *Transactions of the Royal Institution of Naval Architects* (RINA), Vol. 134, 1992.
14. Caprio, F. and Pensa, C., 'Experimental investigation on two displacement catamarans: Systematic variation of displacement, clearance and stagger', *International Journal of Small Craft Technology*, RINA, 2007
15. Jamaluddin, A., Utama I.K.A.P. and Molland, A.F., 'Experimental investigation into the drag characteristics of symmetrical and asymmetrical staggered and unstaggered catamarans', *ICSOT Indonesia 2010*, RINA International Series Conference, Surabaya, Indonesia, 11-12 November, 2010
16. Jamaluddin, A., Utama, I.K.A.P., Aryawan, W.D., Widodo, B., 'Experimental Investigations into the Resistance Components of Symmetrical Catamarans with Variation Hull Clearances and Staggers', *International Journal of Small Craft Technology*, Vol. 154, Part B1, Transactions of RINA, June, 2012.
17. Utama, I.K.A.P., Jamaluddin, A., Aryawan, W.D., 'Experimental Investigations into the Drag Interference of Symmetrical and Asymmetrical Staggered and Unstaggered Catamarans', *Journal of Ocean Technology*, Vol.7 No.1, 2012. pp. 42-52.
18. Jamaluddin, A., Utama, I.K.A.P., Aryawan, W.D., Widodo, B., 'Experimental Investigations into the Resistance Components of Asymmetrical Catamarans with Variation Hull Clearances and Staggers', *International Journal of Small Craft*

Technology, Vol. 154, Part B1, Transactions of RINA, June, 2012.

6. AUTHORS BIOGRAPHY

Dr. Ir. Andi Jamaluddin, M.Sc. is a senior researcher at the Indonesian Hydrodynamics Laboratory (IHL). He received his M.Sc. in Marine Technology from the University of Strathclyde, Glasgow, U.K. in 1991 and his PhD in 2012 in Naval Architecture from ITS. He has extensive experience with various research areas of ships and offshore structures.

I Ketut Aria Pria Utama, M.Sc, Ph.D, CEng, FRINA is Professor of Ship Hydrodynamics and Head of the Department of Naval Architecture and Shipbuilding Engineering, ITS Surabaya. Prof. Utama received his M.Sc in 1996 and Ph.D in 1999 from the University of Southampton, UK. He has extensive research areas in ship design and hydrodynamics. He is presently the Secretary of RINA Indonesian Branch.

DESIGN DEVELOPMENT AND CONSTRUCTION OF AIR SUPPORTED FAST CREW AND OFFSHORE SUPPORT VESSELS.

U. S Tudem, T. K Livgard, Effect Ships International AS, Norway
E. Taniputra, S. Riyadi, Orela Shipyard, Indonesia

SUMMARY

In response to increasing global market request for more efficient, - economical in use, -capable and – environmental friendlier fast crew boats / offshore support vessels, Effect Ships International AS (Norway) has developed a new Air Supported Vessel (ASV) concept. The startup vessel, a 22 m ASV Mono fast crew boat / multipurpose support vessel, developed in collaboration with Orela Shipyard (Indonesia) will be presented. The new vessel will have twin water jets and 40 + knots performance, and is designed for waters with high density of debris as can be the case in Indonesian waters. The new ASV Mono hull concept is the first in a series of new safe, efficient and reliable ASV work boats, suitable sizes between 15 and 100 m +.

A radical reduction in hull resistance compared with conventional hull forms has been documented. Tank testing at SSPA shows reductions between 30 and 50 percent. Up to 80% of the displacement of the vessel is supported on a cushion of pressurized air, reducing wetted surfaces to a minimum and giving the concept additional benefits in terms of reduced pitch and roll motion leading to improved onboard comfort in a sea state. The air cushion system acts as a very efficient motion damping device.

For vessels operated several thousand hours a year the benefits from ASV on fuel savings, improved performance and transport efficiency will play an increasingly important role for the economy of the operation.
Similar ASV hull forms may also be successfully used on a wide range of other applications, including fast ferries, patrol vessels, various work boats as well as pleasure boats and yachts. This paper will discuss the design development, tank testing results on ASV's and construction of the ASV fast crew boat in Indonesia. The paper will also show some results from a couple of other ASV Mono projects, including a 20 m full size test and concept demonstrator, which contributed to ESI winning the *European Power Boat of the Year 2011 Innovation Award* having demonstrated a 50% reduction in fuel consumption compared with conventional planning hulls.

The new range of ASV's may be fitted with alternative propulsions; included but not limited to: Water Jets, pod propulsion (i.e. Volvo Penta IPS or Rolls Royce pod drives), Servo Gear units and more.
The ESI ASV technology and solutions have been successfully patented in 60 countries worldwide.

1. INTRODUCTION

For fast and medium speed commercial vessels and work boats of various kinds operating at relatively high rating, for 2.000 – 3.000 hours or more pr year, almost any reduction in fuel consumption will be very important for the owner / operator. All over the world fuel related cost is worrying the owners/operators these days, and their willingness to look outside the box for new innovative solutions is rapidly picking up.

It took Eka Taniputra, managing director of Orela Shipyard, less than 20 minutes testing with Effect Ship's ASV Mono concept demonstrator in Turkey to be convinced on the new concept's efficiency, performance and potential. Orela Shipyard decided to set up a licensing / collaboration agreement with ESI and to proceed to building of an Indonesian ASV demonstrator. While ESI's demonstrator had small Volvo Penta IPS propulsion and was primarily a large volume pleasure boat type of design, the Orela vessel would use a new ASV Soft Motion hull and WJ propulsion. The Orela ASV will be specified and outfitted as a fast crew shuttle vessel, but will also demonstrate use of the concept for a wide range of other applications.

2. ASV CONCEPT BACKGROUND

ESI has conducted research and technology development on air assisted hulls for more than a decade. More than Euro 12 million has been invested in the developments. The ASV technology may be used on mono- and multihulls. ASV catamaran hulls for fast ferries have been extensively developed and tested. More than 10 different tank testing models have been designed, outfitted and tested at the SSPA Sweden towing tank in Gothenburg. Testing has been carried out in calm waters (hull resistance and ASV systems / lift fan powering investigations), and testing of onboard motions/comfort aspects, added resistance and general behavior in head- and following waves. Applications have primarily been fast ferries and RO-

RO ROPAX vessels. The 40 meter size vessels were typically tested for speeds all the way up to 70 knots, while 125 m ROPAX vessels were tested for speeds between 40 and 60 knots. These designs were set up for WJ propulsion.

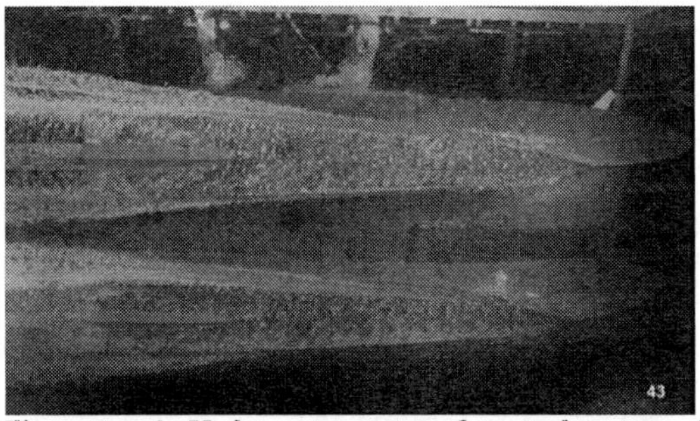

Illustration 1: Under water picture from tank testing of 70 knots 40 m ASV catamaran ferry design.

Illustration 2: Underwater picture from tank testing of 60 knots 125 x 45 m ASV cat ROPAX design.

Very encouraging results on efficiency were documented, particularly for the higher speeds. The 40 m ASV cat designs tested during 2002 - 2005 exhibited an overall power reduction of approx 30 – 40 % at 55 knots. ESI also developed manned models of the two designs and carried out very encouraging tests with these 10 – 11 m demonstrators.

Illustration 3.ASV catamaran 40 m fast ferry design – 11 m manned model with WJ propulsion.

Illustration 4. ASV catamaran 125 m ROPAX design, testing.

In the early 2000 the market for fast passenger ferries changed quite dramatically. Whereas speed used to be highly requested and fuel conception was modestly important; the drastic rise in fuel prices combined with emerging interest and concern for environmental impact from fast ferries, changed this sector. To obtain more acceptable fuel consumption levels speed started to drop. The so called fast ferries were throttled back and 25 – 30 knots was the new norm.

During 2004 – 2006 ESI started to work on ASV Mono hull forms. Both large and fast ASV Monohulls with WJ for fast express cargo, military applications etc, and smaller ASV designs for pleasure boats were developed. Instead of first going through tank testing, ESI considered all the research and results from the ASV catamarans to be sufficient to proceed directly to building a full scale ASV Mono test and documentation vessel.

3. DEVELOPMENT OF THE ASV MONO DEMONSTRATOR

It was decided to build a 65' (20 m) ASV mono test- and demonstration vessel, with a generous beam of 5.4 m, featuring a large volume pleasure boat type of hull. Whereas comparable vessels of this size typically have 2 x 1200 – 1360 HP installed, the ASV should have modest Volvo Penta IPS 600 pod drives with 2 x 425 HP propulsion power. Design speed was set to be 30 knots. In addition a separate smaller engine was used to power the lift fan system, via a Bosch Rexroth hydraulic system.

3.1. HULL FORM

The shape of an ASV Mono resembles a catamaran almost as much as a monohull. On either side of the air cushion chamber a propulsion gondola is housing the engine/driveline/propulsor. The lower part of these bodies form relatively short planing surfaces contributing to dynamic lift already at relatively low speeds.

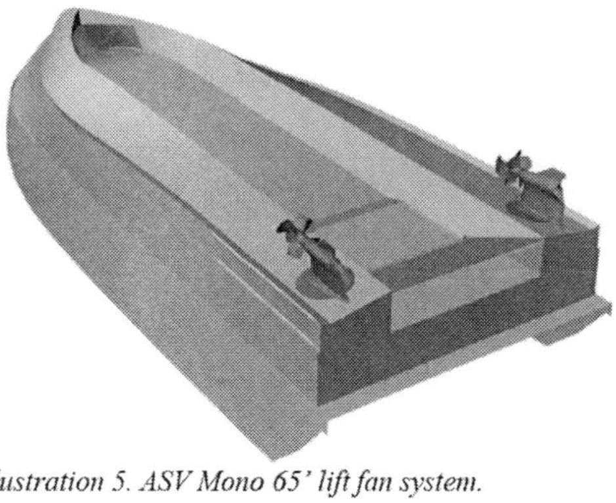

Illustration 5. ASV Mono 65' lift fan system.

In the bow of the vessel a double inlet, single outlet

centrifugal fan is located. The air to the fan comes through vanes on either side of the upper forward part of the hull. The pressurized air from the fan is fed into the air cushion chamber through a duct. Illustration 5 is showing the air cushion chamber with the air inlet to the chamber forward. The ceiling of chamber is normally submerged in water when the vessel is off the air cushion. The capacity of the lift fan is up to 15 m3 / sec of air with a pressure of approx 4,8 kPA. During normal operation in relatively calm waters only a fraction of this capacity is used.

Illustration 6. ASV Mono 65' lift fan system.

Illustration 7. ASV Mono 65' from inside the 4 kPa pressurized air cushion chamber.

The aft part of the air cushion is sealed with a proprietary air cushion enclosure flap arrangement. No rubber skirts or other high maintenance systems are used.

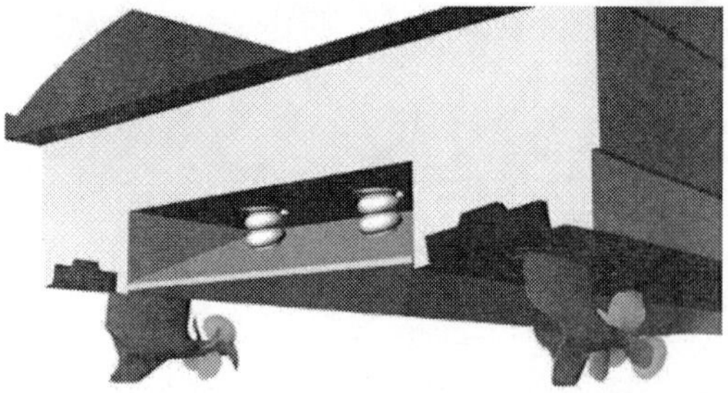

Illustration 8. ASV Mono 65' showing IPS propulsion and air cushion enclosure flap arrangement.

3.2. EFFICIENCY AND TEST RESULTS

The air cushion support rate for the ASV Mono 65' is typically between 55 and 65 %. The vessel rides throughout the whole speed range with almost neutral trim. A modest ½ to 1 degree bow up is typical. The air cushion is effectively sealed as the full air cushion periphery is submerged to closely the same level. With almost vertical hull sides the vessel has a larger than normal water foot print, resulting in a "light" and dry ride. With the lift fan system engaged, the vessel lifts, already from stand still and without any dynamic support from planing surfaces, directly up some 40 cm. The acceleration is fast and very efficient up to cruising speeds without the typical hump to overcome. The ASV Mono 65' is fully instrumented, measuring among other fuel consumption vs. speed and torque in order to have a proper understanding on total power consumption (for propulsion and lift fan combined) at different speeds. The results of the ASV and compared with other vessels of the same category is presented in the illustration 9 graph.

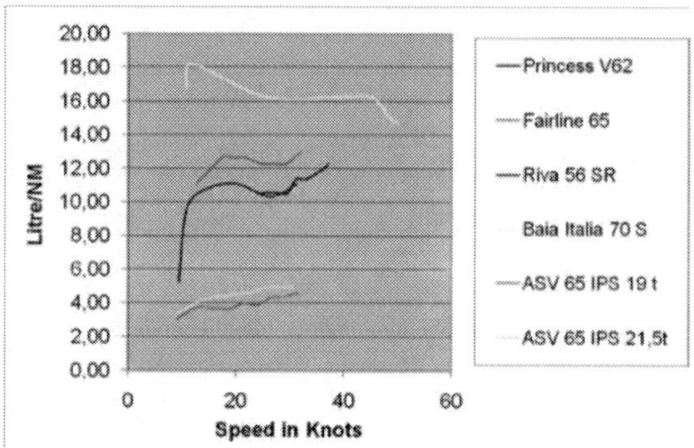

Illustration 9. ASV Mono 65' fuel consumption compared with other conventional vessels of same size.

In short the ASV is consuming only about 50% of the fuel the comparable vessels do, and even more importantly the advantage is over a very wide range of speeds; already at 10 – 12 knots in the 20 m scale.

Consumption values of less than 5 L/Nm at 25 – 30 knots for a vessel of this size is quite remarkable.

Depending on loading the ASV Mono 65'achieved a top speed of nearly 38 knots at 16 tons, 32,6 knots at 21- 22 tons and 31,6 knots at 26,5 tons.

Illustration 10. The ASV Mono 65' underway at 30 + knots.

Wash wake conditions behind the ASV is equally impressive. The much reduced wash wake illustrates how the ASV concept is using the power to propel the vessel efficiently forward and not to produce large and unpleasant waves.

Fan powering is dependent on operational conditions. The air cushion system is designed to reduce wetted surface to a very low level, and to reduce air flow / ventilation of air from the air cushion. The less air ventilated from the cushion the less power needed to keep the vessel "cushion borne". The fan system is specified to have a good air flow (m3/sec) margin; to be able to handle more ventilation typically occurring during operations in a sea state. Even at relatively high and short waves for vessel size the ASV copes without "falling off the cushion" and/or having water slamming through the air cushion and up into the cushion ceiling, The air cushion is when operated in waves acting as a "motion damping device", contributing efficiently to improve onboard motions.

4. TANK TESTING OF ASV MONO HULL FOR A FAST BATTERY POWERED COMMUTER FERRY

During the winter 2012 season ESI carried out tank testing of another ASV Mono hull, to be used for a European Union (EU) sponsored project dedicated to developing the world's first zero emission and 30 + knots fast battery powered commuter ferry for sheltered waters.

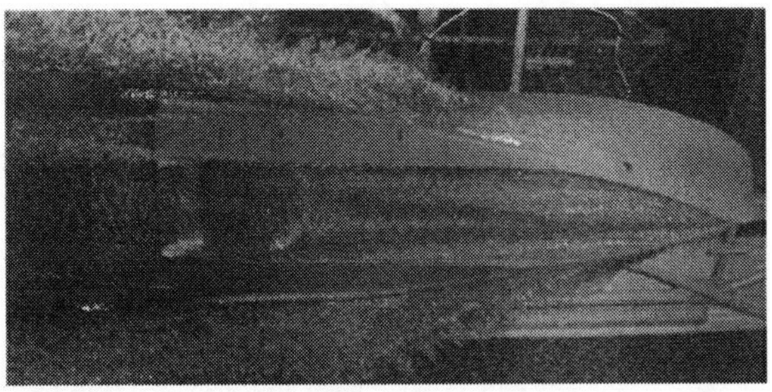

Illustration 11. Underwater picture of ASV Mono BB GREEN hull form, tank tested at SSPA during winter 2012.

Although not the "normal" sequence for concept developments/verification, the BB GREEN Mono tank model, demonstrated excellent efficiency fully in line with the data already available from the ASV Mono 65' test vessel. As part of this project also a similar size ASV Catamaran was tested to verify which of the two – the ASV mono or the ASV cat – would perform the best. In illustration 12 the hull resistance of the two ASV's is compared with State of the Art conventional planing hulls with equal design speed (35 knots). The results confirm the performance already seen on the full scale vessel and are very important for verification of scaling effects for evaluation of ASV concepts also for other vessel sizes.

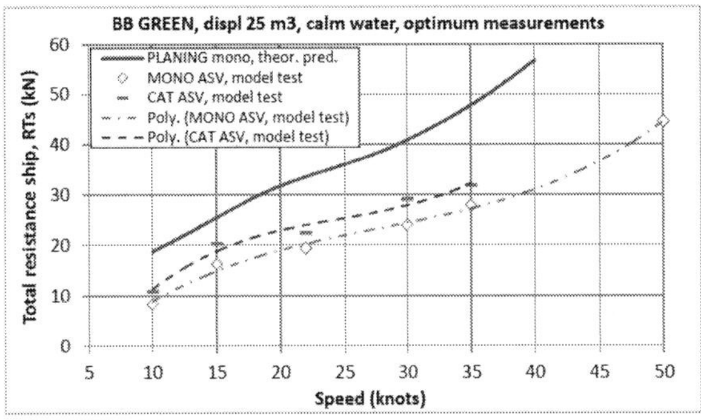

Illustration 12. ASV Mono BB GREEN hull resistance.

Although both the ASV designs exhibit much lower hull resistance than conventional "wet" vessels, the ASV Mono comes out at the top with impressive values throughout the whole speed range.

Another finding from the ASV Mono 65' test vessel documented in the towing tank is the very low wash wake (energy) of the ASV Mono. In below illustration a number of projects have been compared. The ASV Mono BB GREEN comes out clearly at the top. Low wake wash is a very positive feature for operating a fast commuter vessel in densely trafficked waterways.

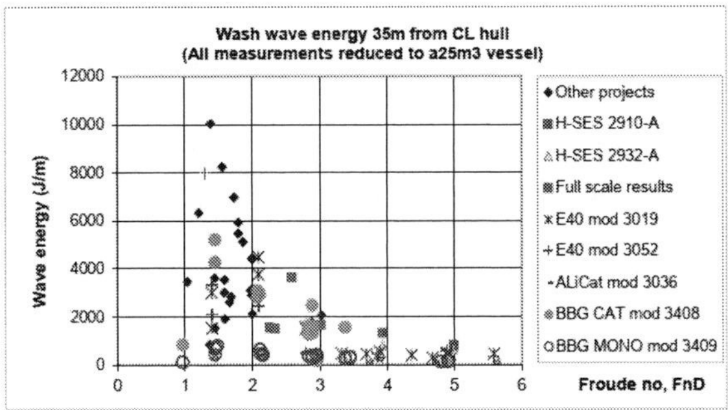

Illustration 13. ASV Mono BB GREEN

5. ASV MONO SOFT MOTION HULL FORMS

ESI has earlier and with support from EU successfully developed and tested ASV catamarans with Soft Motion (SM) characteristics. The idea to the ASV Mono SM was uncovered when US Navy tested an earlier ASV cat demonstrator; was very pleased with the motion damping of the air cushion and asked if the excellent ride could be replicated on ASV Mono hulls.

Although the ASV mono from the outside looks very similar to a conventional mono hull, the vessel is in reality a mix between a catamaran and a monohull, where air support and reduced wetted surface areas are combined with displacing bodies and planing surfaces; contributing to extra "lift" at speed.

When at rest and in "off cushion mode" the ASV will, in the same fashion as a catamaran, have self righting and roll suppressing properties. At speed the vessel has been designed to operate at close to neutral trim. Minor trim adjustments can be made by means of interceptors (Humphree) mounted on each of the propulsion gondolas at the stern.

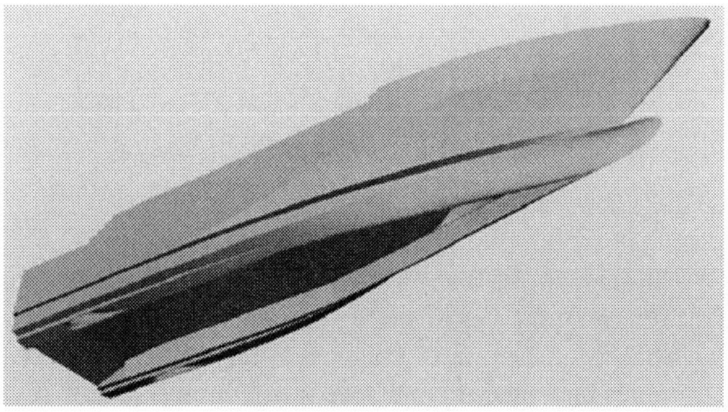

Illustration 14. ASV Soft Motion hull with WJ propulsion.

Generally speaking the ASV Mono can be configured for most known propulsion systems; water jets, pod propulsion, Servo Gear variable pitch propellers, stern drives or conventional shaft propellers.

The innovative ASV Mono Soft Motion hull is characterized by a sharp bow section, with a narrow bulbous bow, typically stretched out to the tip of the bow to take advantage of increased water line length. Most of this bow will remain out of the water when the vessel is travelling at high speed in calm waters, and will be acting as a combined volume body / pitch damping unit when travelling at speed in waves.

Above this wave piercing bow, the lower part of the hull is quite narrow and sharp at first, widening out to meet a knuckle line – arranged quite high at the utmost bow and gradually dropping until it meets the wide spray rails where the hull widens out. The purpose is to reduce accelerations resulted from fast wave impacts.
An ASV's "water foot print area" is mainly taken up by a V-shaped air cushion / cavity, starting directly behind the bulbous bow and running the full length of the hull until meeting the air cushion enclosure flap. The main function of the hinged flap is to maintain the pressurized air inside the air cushion chamber, and does not have a trim tab function. The flap and lift fan system is controlled by an atomized management system with overriding options if the master has a different preference to the standard settings.

The lift fan system /centrifugal fan, is located in the bow of the vessel, and is powered either by hydraulic or electric motor to deliver sufficient pressurized air to the air cushion chamber, and to maintain a steady pressure inside the cavity even in demanding sea states.

The propulsion gondolas, typically one on each side, of the hull have several functions; housing the propulsion system, securing buoyancy (as pontoons) and acting as dynamic/planing surfaces at speed. Due to the relative short length of these bodies and the angle of attack between the gondola and the water, the lift is considerable already at relatively low speed. For a 20 m vessel the planing part of the gondola can be approx 4 – 5 m, and will contributing with planing effect at approx half the speed of the main vessel.

The forward part of the propulsion gondolas/side keels is quite narrow with a deep V shape. The main function is to enclose the air cushion well to avoid access ventilation from the air cavity; with a minimum of wetted surface areas. The sharpness of these hull elements is also positive from a motion point of view when operated at high speeds in waves. Heavy pitching of the bow will increase ventilation, fan power and overall resistance.
Moving aft wards the gondolas widens into planing surfaces. The location - around the LCG, - will contribute to a soft heave support, contradictory to a conventional planing boat where the planing surfaces more often are located far forward (and sometimes also aft) typically generating uncomfortable pitch motions.

The air support rate can be varied but is typically between 50 and 80 % depending on hull configuration/loading etc.

6. ORELA ASV MONO SOFT MOTION CREW BOAT

As part of collaboration between ESI and Orela, Orela is building an ASV Mono crew boat with the following main characteristics:

Hull length	22,0 m
Hull beam	5,3 m
Construction material	Aluminum
Operational weight full load	Approx 40 tons
Propulsion	2 x MJP water jets 450 CSU
Main engines	2 x 880 kW Doosan (MAN)
Lift fan powering / auxiliary fire pump engine	1 x 240 kW
Lift fan system	Centrifugal fan
Air cushion area	Approx 46 m2
Air cushion pressure	Approx 5,5 – 7 kPa
ASV management system	ESI
Est. design speed light	45 knots +
Est. speed design loaded	40 knots +
Seating capacity with cargo	50 Pax on main deck

Seating with crew module	90 Pax on main deck
Aft deck free area	Approx 40 m2

The Orela ASV demonstrator design is showing in illustration 15, and from construction illustration 16.

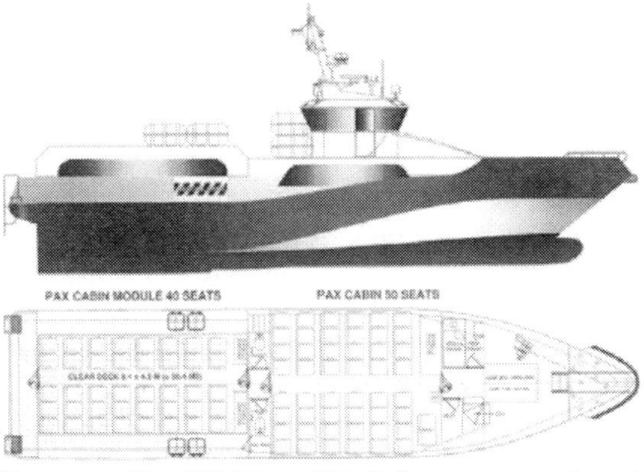

Illustration 15. ASV Mono SM 40 knots + crew boat / concept demonstrator (Orela Shipyards, Indonesia).

Illustration 16. From construction of ASV Mono crew boat.

Propulsion for this vessel will consist of twin MJP 450 CSU (Compact Steering Units) water jets with specially designed water jet intake / water jet pick up geometry. With a propulsion power of approx 2 x 880 kW the vessel is estimated to achieve nearly 40 knots at full load.

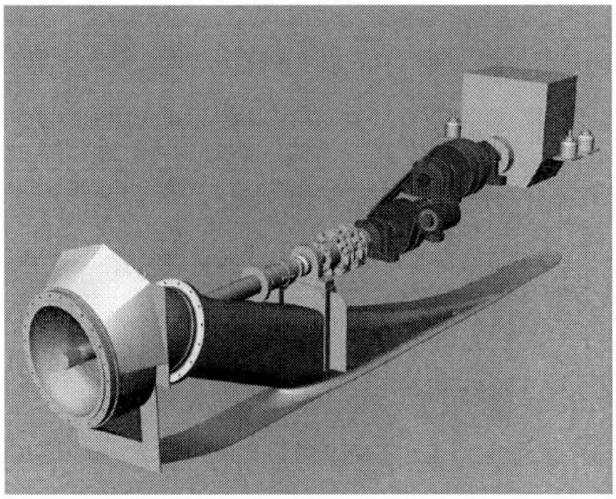

Illustration 17. Aft end of ASV Mono SM drive line.

For ease of installation and maintenance the power from the main engine/gearbox to the specially designed water jet is shifted vertically and horizontally through a Polychain arrangement, showing in illustration 17.

The fist vessel will have seating capacity for 50 Pax on the main deck. Aft of the cabin area there will be a large deck area of almost 40 m2 to take different type of light cargo, container modules of different kinds or as showing on the general arrangement (GA), a second passenger cabin module with additional 40 seats. In addition space is allocated for crew accommodation, toilets, medical room, galley, baggage storage etc. Other general arrangements will naturally be on request depending on application / end users requests.

The purpose of the ASV vessel is through demonstration under actual operational conditions to prove feasibility of new concept. The new vessel will be used by the yard to demonstrate functional (handling, comfort), economical and environmental benefits with the ASV concept compared with current conventional vessels. Although the vessel is set up as a fast crew boat and configured for the Indonesian offshore market, it will also be used to demonstrate and document potential use of the same new hull concept for other applications and markets.

Construction of the hull and installations of the systems are in progress at the Indonesian yard, and planned launch for initial testing and trials is early 2013. Following completion and documentation of performance and capabilities of the new ASV, the vessel will be used extensively in generating awareness of the new environmental friendly concept and for marketing and sales purpose. Interested parties are encouraged to contact either ESI or Orela for information on construction progress, availability of test results and for setting up an appointment for testing the new concept at sea.

As information on the new vessel type is starting to spread, Effect Ships and Orela Shipyards are already reporting considerable interest from potential ship owners and operators.
Other yards are also closely following the developments.

Based upon feedback from the market, Orela has already started the design of larger ASV's, among other 34-36 m vessels for various commercial and navy/paramilitary applications.

Fast and efficient passenger ferries with ASV Mono technology seem to be yet another likely application.

6. ASV SOFT MOTION TANK TESTING

In Scandinavia ESI has another ASV Soft Motion project ongoing. Partly funded by Innovation Norway as a maritime environmental project, with a Norwegian leading coastal ferry operator and a yard specializing in light weight carbon sandwich construction as partners; development of a new range of fast, efficient and capable coastal ferries are ongoing. The ferry operator sees potential to save 30 % or more on his overall fuel bill and to achieve improved onboard comfort for the passengers and crew.

For verification of the hull resistance / powering, motions in waves and other basic assessments, the project will yet again use the SSPA Sweden towing tank in Gothenburg. A full week of testing will be completed during week 43, and according to plan the first results and findings will be presented at the conference.

As will be seen from below illustration, the hull form resembles closely to the Orela work boat design, but is configured for Volvo Penta IPS or any alternative pod propulsion.

Illustration 18. ASV Mono SM tank testing model for fast ferry and work boat applications to be tested at SSPA during week 43 2012. A few design elements not showings.

Italian designer Mauro Sculli will be responsible for the design of this new ferry range covering passenger only and passenger/light cargo designs from 20 m to 40 m +. Naturally these designs may also be used for aluminum construction.

7. ALTERNATIVE PROPULSION SET UPS FOR ASV

The ASV Mono hulls may be configured for a wide range of propulsion systems.

Systems include:

- Volvo Penta IPS
- Rolls Royce new carbon pod propulsor
- ZF / Zeus range of pod propulsors
- Single pod proulsion (ZF) with diesel/electric hybrid set up. A special ASV hull form will be used
- Servo Gear variable pitch propulsors
- Various brands – stern drives (for smaller vessels)
- MJP and Rolls Royce Water jets
- Surface piercing prop systems (i.e. Arneson)
- Conventional fixed pitch propulsors

Owner/operator preferences, operational / use profile and conditions, operational speed range, local conditions (i.e. Indonesian problem with a lot of debris in the water) are among the factors to be considered prior to selection and integration of propulsion for the targeted ASV project.

7. EXPLOITATION OF ASV TECHNOLOGY AND SOLUTIONS.

The proprietary ASV technology has been developed and is owned by ESI. Licensing and collaboration agreements will be established with selected yards and other parties around the world. These companies will through payment of a license fee, be given access to use the latest in ESI ASV technology hull forms and solutions for construction of various kinds of air supported vessels according to yard's and/or end users requirements and topside designs.

8. CONCLUSIONS

With current very high fuel prices and no expectation for lasting drops, particularly vessels and operations with a high number of operational hours pr year will benefit strongly from use of the new ASV technology.

With markedly less consumption and ability to offer higher speeds without jeopardizing diesel expenditures ASVs will offer improved transport efficiency and economy in use. On a global basis there is today a strong market demand for more efficient and economical vessels. Effect Ships International has a wide range of ASV hull forms to suit most owner/operator's requirements.

Documented results from tank testing as well as from full scale ASV Mono hulls; highlighted through comparison studied with market leading conventional hull forms, conclude very positively in the favour of the ASV Mono concepts.

Due to massive reduction in wetted surface area and corresponding hull/water resistance - in several cases by approximately 50 % - these new ASV hulls seem capable of accomplishing a long overdue technological shift. The new concept also benefits from its simplicity and robustness and lack of "old fashion" rubber seals or skirts as seen on conventional surface effect ships (SES), hovercrafts or hybrid SES solutions. The power to "stay on the cushion" for an ASV mono is low, in calm waters representing only 3-

5% of installed power, in a sea state naturally somewhat more due to more ventilation / air flow.

Well worth noticing is the fact that great efficiency improvement does not only occur at high speeds. The ASV Mono offers superior efficiency a very wide range of speeds; from below 10 knots all the way to 50 knots + (exemplified for a 20 m vessel), making the concept suited for a wider range of vessel sizes and speeds.

The air cushion system offers motion damping in waves. Wave impact, pitch and roll motions are reduced not only at speed but also at rest in waves. The ASV Mono SM will be particularly suited for operation in more demanding waves/sea states; conditions many work boats are frequently encountering.

From a construction point of view an ASV Mono hull is generally simpler and less expensive to build; demands less material with lower weight – compared with a similar size catamaran. As for any medium/high speed vessel lower weight gives less power requirement. Aluminium construction, GRP or carbon sandwich are recommended.

Unlike slender catamarans, the interior volume and space an ASV Mono can offer is excellent. Typically the engine/driveline size is smaller than for a conventional vessel, the sides of the hull are close to vertical and the fan system takes up little space. Most of the volumes above the air cushion ceiling can therefore be used for accommodation etc.

8. REFERENCES

ALLENSTROEM, B, LILJENBERG, H, TUDEM U. *"Concept development and model testing - new generation Air Assisted Vessels (AAV) with Water Jet propulsion"*, FAST 03, Ischia, Oct 2003

EILERTSEN, A, GRIEG EIDE, J, TUDEM, U. *"Design development of advanced air supported Vessel (ASV) designs. "* RINA WARSHIP, London June 2004.

LIVGARD, T K, TUDEM U. *"High Performance with Improved Onboard Comfort: Introducing ASV "Soft Motion" Mono With Water Jet Propulsion, RINA SURV 7, Pool, UK, May 2009*

LIVGARD, T K, TUDEM U. *"Design Development, prototype construction and test outfitting of the Air Supported Monohull by Effect Ships International AS". RINA, London Nov 2009*

LIVGARD, T K, TUDEM U. *"Air Supported Vessel (ASV) monohull, design development of test vessel and proof of concept". 31. Symposium Yachtentwurfund Yachtbau, Hamburg, November 2010*

LIVGARD, T K, TUDEM U *"Air Supported Vessel (ASV) monohull demonstrator. Design development, testing at sea and results proving the concept", Naples May 2011*

TUDEM, U *"Air Supported Vessel (ASV) – innovative fast vessel technology with great operational- and environmental benefits". TEKNA 2011 Copenhagen, December 2011.*

TUDEM, U, LIVGARD, T K and LILJENBERG H. *"BB GREEN – development of the world's first Zero emission air supported battery electric fast commuter ferry, concept and first project results". HIPER 2012 Duisburg, Germany. September 2012*

TUDEM, U, LIVGARD, T K and LILJENBERG H. *"Design development of Air Supported Vessel (ASV) hull forms for passenger ferries and work boat applications". HIPER 2012 Duisburg, Germany. September 2012*

ALOE VERA AS NATURAL CORROSION INHIBITOR ON MILD IN SEAWATER

F. Manuhutu, W.B.Wan Nik., M.F.Ahmad, Noor, C.W.M and A.Fitriadhy Universiti Malaysia Terengganu, Malaysia

SUMMARY

Mild steel is mostly used in marine environment where it corrodes very rapidly. This study reveals that aloe vera gel as natural corrosion inhibitor could decrease the corrosion rate of mild steel in seawater. Three different concentrations of aloe vera gel that mixed with 10 litre of seawater taken from the South China Sea and were prepared to produce four different solutions. First is controlled seawater, second is 300ppm concentration of aloe vera gel, third is 600ppm concentration of aloe vera gel and fourth is 900ppm concentration of aloe vera gel. Two methods is used to analyze the corrosion process of steel coupon samples: Weight Loss and Electrochemical Impendence Spectroscopy (EIS) method. The results reveal that high aloe vera concentration enhances the corrosion inhibition efficiency of mild steel in seawater.

NOMENCLATURE

W	Weight loss of mild steel coupon (g)
W_i	Initial weight of mild steel coupon (g)
W_f	Weight of mild steel coupon after immersed in seawater for certain time (g)
cR	Corrosion rate of mild steel coupon (mm/year)
A	Area of one mild steel coupon
T	Time of immersion (hours)
IE	Inhibition efficiency (%)
CR_o	Corrosion rate of mild steel coupon without inhibitor [mm/year]
CR_i	Corrosion rate of mild steel coupon with inhibitor [mm/year]
I_{corr}	Corrosion current (mA)
E_{corr}	Corrosion potential (mV)

1. INTRODUCTION

Uses of mild steel extend over almost all forms of industrial applications and industrial manufacturing due to low cost but this kind of steel must be protected from corrosion to avoid the soaring of the maintenance cost.

It has been found that one of the best methods of protecting mild steel against corrosion involves the use of inhibitors which substances slow down the rate of corrosion. An inhibitor is usually added in small amount in order to slow down the rate of corrosion through the mechanism of adsorption [1].

The natural inhibitors (green inhibitors) are significant because they are non toxic and do not contain heavy metals hence they are environmentally friendly [2]. Natural aloe vera compounds offer interesting possibilities for corrosion inhibition and the particular interest because of their safe use and high solubility in water. The adsorption of the surfactant on the metal surface can markedly change the corrosion-resistance property of the metal, and so the study of the relations between the adsorption and corrosion inhibition is importance.

Marine corrosion usually occurs due to the exposure of metal to marine environment such as seawater. Some of the systems such as ships, marinas, pipelines, offshore structures, desalination plants, and heat exchanger are the examples that experiences marine corrosions. This corrosion problem that affects most of the systems has been well studied for a long time but failure of material in seawater is still occurs [3][4]. The marine corrosion problem produces less desirable material from the original metal and resulted in the reduced of quality of the metal itself to be used in the industry. Most of the material that have been use as corrosion inhibitor nowadays are inorganic substances such phosphate, chromates, dichromate and arsenates. Even it can be an effective as corrosion inhibitor for metal; its toxicity behaviour gives negative effects to the environment [5][6][7]. In the other hand, using chemical substances to act as metal corrosion inhibitor is not only harm the environment but also consume high amount of cost. The use of natural products for metals protection is of much importance because in addition to being inexpensive, readily available and renewable sources of material, plant are environmentally friendly and ecologically acceptable.

2. EXPERIMENT

2.1 MATERIAL PREPARATION

Mild steel coupons with dimension 25mm x 25mm x 3mm is prepared for the experiment. Each coupon was perforated with a hole of the same of about 1 mm diameter at one end to allow the passage for hanging the sample during immersion process. Before exposure, more than 100 samples were polished manually using 500, 600 and 800 grit of emery papers. The samples then cleaned by acetone and then rinsed with distilled water to avoid any reaction occur on the coupons. Is prepared three different seawater solution with concentration of aloe vera 300ppm, 600ppm and 900ppm is prepared using seawater collected from South China Sea along the coastal area of University Malaysia Terengganu.

The solutions were kept in different containers A, B, C and D for the immersion process during 50 days. Container A contained controlled seawater, container B seawater with 300ppm aloe vera gel concentration, container C seawater with 600ppm concentration of aloe vera and container D seawater with aloe vera concentration 900ppm.

2.2 WEIGHT LOSS MEASUREMENT

Weight loss method is used to determine the loss of weight of the mild steel in a period of immersion time in seawater using formula (1).

$$W = W_i - W_f \quad (1)$$

Where: W, W_i, and W_f are the weight loss, sample initial weight and sample present weight respectively.

The weight loss W was calculated in g as the difference between the initial weight W_i prior to immersion and weight after removal of the corrosion product W_f.

2.3 ELECTROCHEMICAL MEASUREMENT

Autolab software was used for electrochemical measurement, this device is connected to the computer to get the data. The measurements are called General Purpose Electrochemical System (GPES) and potentiodynamic polarization (PP), the exposed area of the metal is 12.5cm^2 which was immersed in the test solution for measurement. The scanning rate 0.0101Vs^{-1} was set and the electrode potential was automatically changed from -0.5mV to +0.5mV to get the potential curve and potentiodynamic current. The graph obtained corrosion current (I_{corr}), corrosion potential (E_{corr}).

3. RESULT AND DISCUSSION

3.1 WEIGHT LOSS ANALYSIS

The weight loss is a simple method that has been used in this study to gain the corrosion rate mild steel and efficiencies of the inhibitors.

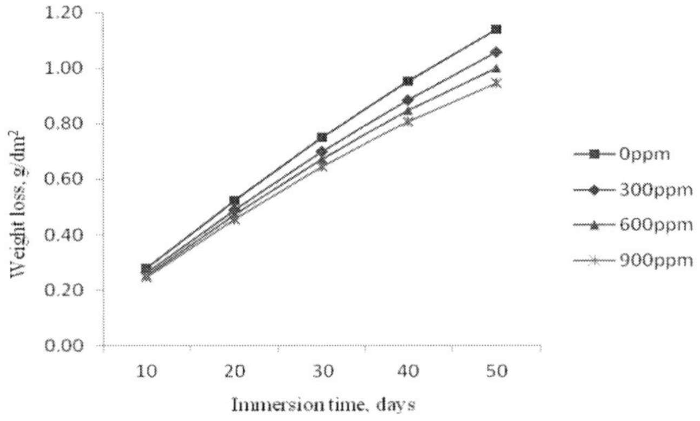

Figure: 1: Weight loss of mild steel in seawater inhibited by Aloe Vera

Fig. 1 shows the graph of weight loss versus immersion time. At first 10 days immersion time, the molecules of aloe vera were started to cover the metal surface to form a thin layer protecting the metal from corrosion, where the film surface that form at 300ppm, is more thin compare to 900ppm. The film formed act as a barrier between metals surface with the seawater as the environment. As thin film formed, it is easy to break and highly exposed to corrosive media compare to film formed at 900ppm concentration of aloe vera. As increasing of immersion time, the weight loss of mild steel at each aloe vera concentration is increased.

Corrosion Rate formula:

$$cR = 87.6 \times W/(DAT) \quad (2)$$

Where,

cR – corrosion rate, mm/year
W – weight loss, mg
D – density of mild steel, g/cm^3
T – time, hour

The increase of aloe vera concentration as inhibitor of mild steel in seawater decrease the corrosion rate, see Fig.2.

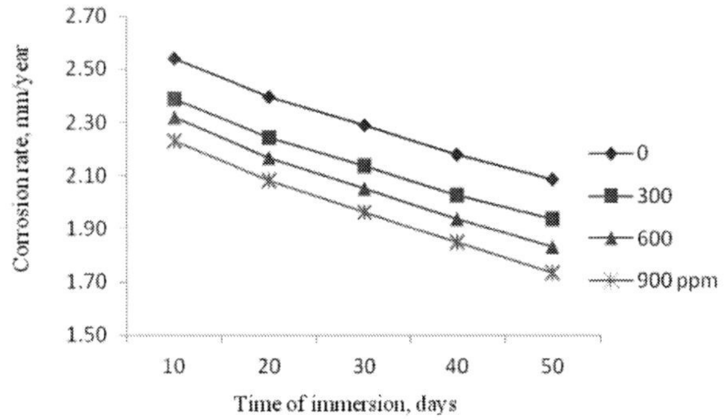

Figure 2: Corrosion Rate of mild steel in seawater with Aloe Vera inhibitor

The inhibition efficiency is calculated by using the formula (3)

$$IE(\%) = \frac{CR_o - CR_i}{CR_o} \times 100 \quad (3)$$

Where, IE , CR_o *and* CR_i, are the inhibition efficiency , corrosion rate without inhibitor and corrosion rate with inhibitor respectively. By the decreasing of the corrosion rate due to increasing of the aloe vera concentration, hence the inhibition efficiency of mild steel corrosion in seawater increases.(Fig. 3).

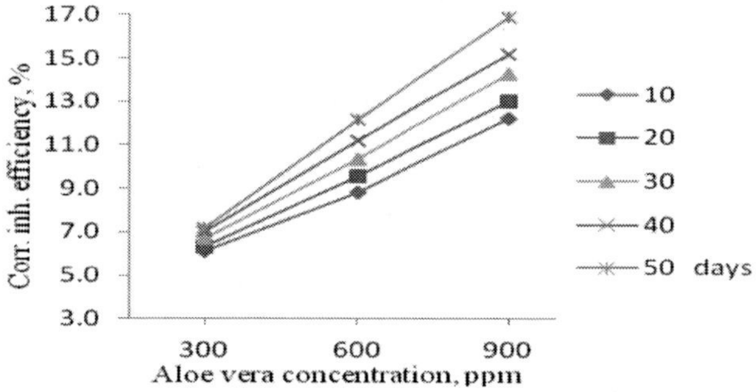

Figure 3: Inhibition Efficiency of Aloe Vera on mild steel corrosion in seawater

3.2 POTENTIODYNAMIC POLARIZATION TEST

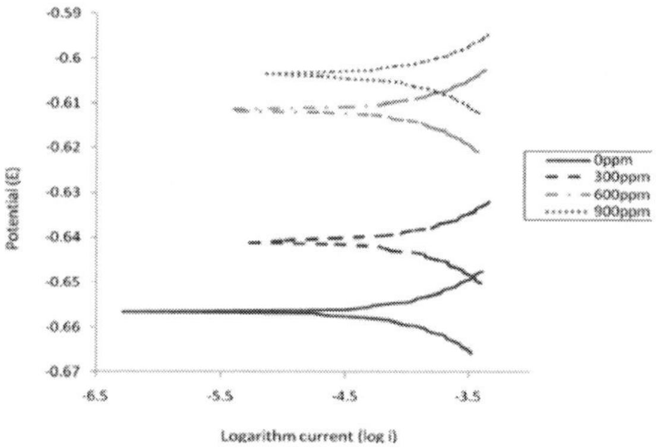

Figure 4: Potentiodynamic polarization cutrves for mild steel
at different concentrations of aloe vera gel.

Fig. 4 shows the potential (E) versus logarithm current (I) graph for 50 days immersion period. The changes observed in the polarization curves after addition of the inhibitor are usually used as criteria to classify inhibitors as cathodic, anodic or mixed. Aloe Vera is categorized as mixed type inhibitor. This can be explained that by the increasing of concentration of Aloe Vera gel the corrosion potential (E_{corr}) is slightly shifted to the positive values direction. This means that increasing of Aloe Vera concentrations has significant effects on the retarding of the cathodic hydrogen evolution reaction and inhibiting the anodic dissolution of mild steel. The potential curve of 0ppm seawater was shifted more to the negative direction compare to the values of the others potential curve that means in the pure seawater the sample had higher corrosion rate than of the other samples [8][9]. The addition of aloe vera gel had reduced the corrosion rate of the samples as their polarization curves are plotted higher than the polarization curve of controlled seawater. In the other hand, inhibition process with 900ppm of aloe vera has the highest corrosion resistance than the others.

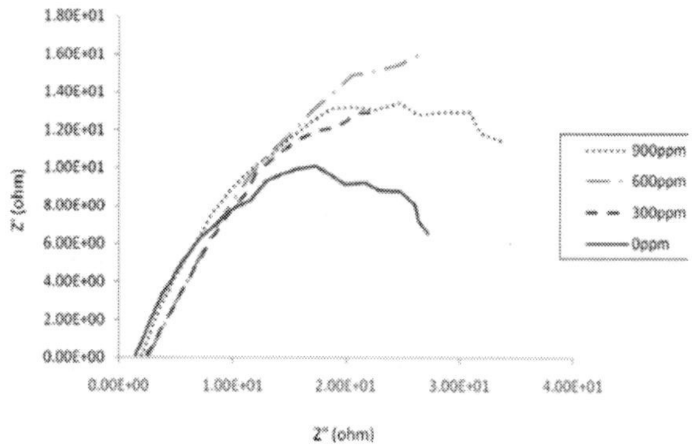

Figure 5: Nyquist plots of mild steel for 50 days of immersion

GPES data for 10 - 50 days of immersion is shown at Fig. 5, the data is presented in Nyquist plots. The x-axis indicates real impedance (Z') while the y-axis indicates imaginary impedance (-Z"). Size of semicircle decreases with decreasing the inhibitor concentration and it indicates a high corrosion rate.

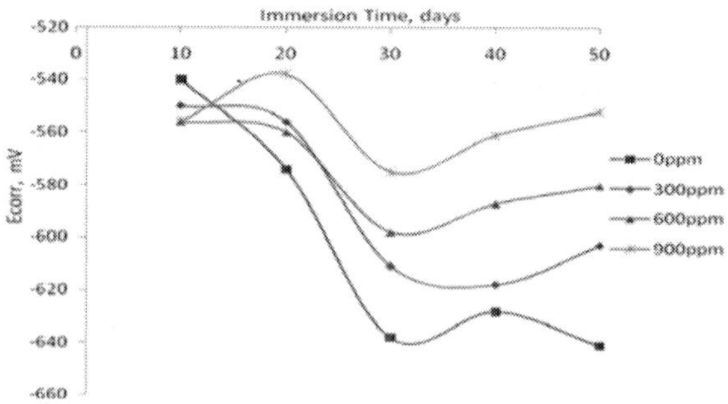

Figure 6: Corrosion potential E_{corr} versus immersion times

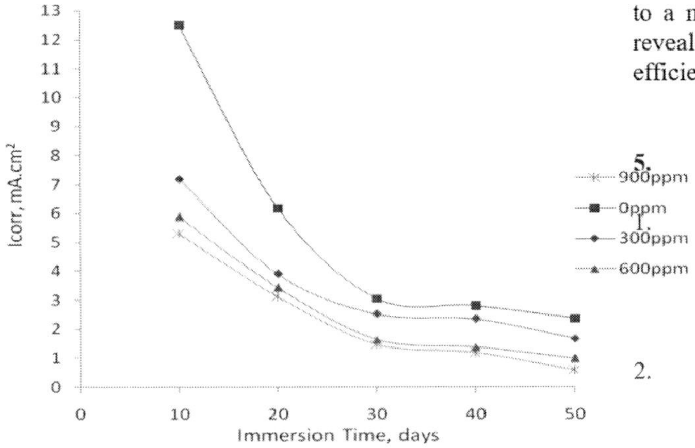

Figure 7: Corrosion current i_{corr} versus immersion time

By the increasing of the aloe vera concentration the values of i_{corr} decreased (Fig. 7) and that lead to the decreasing of the mild steel corrosion rate, it is shown in Fig. 2.
Simultaneously with that the value of E_{corr} at the early period shifted to more negative value, but then shifted to positive values so the system tends to decrease its metal dissolution and hydrogen evolution process (Fig. 6).

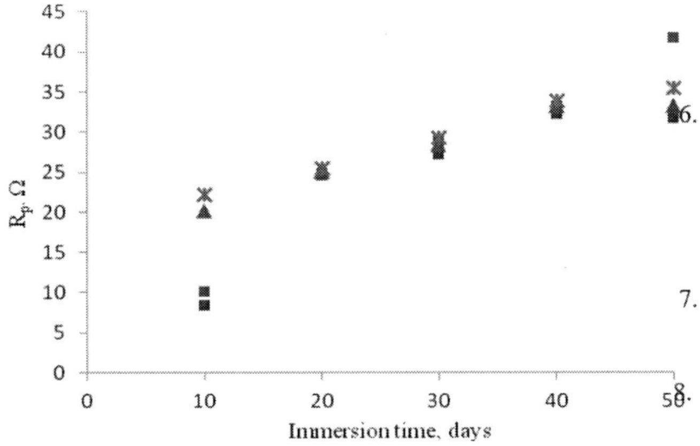

Figure 8: Polarization resistance versus immersion time

R_p values increase due to the incresing of aloe vera concentration indicate that a surface film is formed by adsorption on the surface of mild steel coupon which increase the degree of corrosion protection of mild steel in seawater (Fig. 8).

4. CONCLUSIONS

The present study shows that the increasing of aloe vera concentration as inhibitor on mild steel immersed in seawater decrease the weight loss, the corrosion rate to a moderate degree less than 1.8mm/year, which is revealed by the increasing of the corrosion inhibition efficiency and the polarization resistance.

5. REFERENCES

1. EDDY, N.O., ODOEMELAM, SA. Inhibition of the corrosion of mild steel in H2SO4 by ethanol extract of Aloe vera. *Resin and Pigment Technology* 38(2),2009,111-115.
2. M. SANGEETHA, S. RAJENDRAN, T. S.MUTUMEGALA, A.KRISNAVENI, Green corrosion inhibitors-An Overview. *ZAŠTITA MATERIJALA* 52(1),2011
3. W.B. WAN NIK, F. ZULKIFLI, M.M. RAHMAN, R. ROSLIZA, Corrosion Behavior of Mild Steel in Seawater from Two Different Sites of Kuala Terengganu Coastal Area. *International Journal of Basic & Applied Sciences IJBAS-IJENS,* 11(6), 2011
4. RAMESH, S., RAJESWARI, S., MARUTHAMUTHU, S., Effect of inhibitor and biocide on corrosion control of mild steel in natural aqueous environment. *Materials Letters,* 57, 2003, 4547-4554.
5. ABIOLA,O.K., JAMES, A.O., The effects of Aloe Vera extract on corrosion and kinetics of corrosion process of zinc in HCL solution. *Corrosion science*, 52, 2010, 661-664
6. El-SHAMY, A.M., SOROR, T.Y., EL-DAHAN, H.A., GHAZY, E.A., EWEAS, A.F., Microbial corrosion inhibition of mild steel in salty water environment. *Material Chemistry and Physics,* 114, 2009, 156-159.
7. Robert E. Melchers *, Robert Jeffrey. Early corrosion of mild steel in seawater. *Corrosion Science,* 47, 2005, 1678–1693
8. OBOT, I.B., EGBEDI, N.O.O. Adsorption properties and inhibition of mild steel corrosion in sulphuric acid solution by ketaconazole. *Corrosion Science,* 52, 2010, 198-204
9. JAMES, A.O., OFORKA, N.C., ABIOLA, O.K., ITA, B.I. Corrosion inhibition of mild steel in hydrochloric acid by 2benzoylpyridine and pyridoxolhydrochloride. *E-Journal of Chemistry*, 6(2), 2008, 385-394

6. AUTHORS' BIOGRAPHY

Ferry Manuhutu, contract lecturer at University Malaysia Terengganu(UMT), Malaysia. Lecturer for subjects: Marine Corrosion. Completed research: A study of Allium Cepa as Natural Corrosion Inhibitior on Mild Steel in Seawater. Ongoing research: Influence of flow velocity and angle of attack on mildsteel coupons immerse in seawater. Supervising final year projects of undergraduate students (B.Appl Sc) in marine corrosion: Effect of High Concentration on Corrosion Performance of Aluminium Alloy Inhibited by Henna.

W.B. Wan Nik, Associate professor in Marine Corrosion, dean of Faculty of Maritime Studies and Marine Science UMT, Malaysia. Lecturer in Marine Corrosion. Supervising master and PhD students in field of marine corrosion. Publications in marine corrosion: Potential of Honey as Corrosion Inhibitor for Aluminium Alloy in Seawater, World Applied Sciences Journal, V14, n2, p.215-220, The Corrosion Behavior of Mild Steel in Seawater from Two Different Sites of Kuala Terengganu Coastal Area, International Journal of of Basic & Applied Sciences (IJENS), ISSN 114405-06-7676, V11, N6, p. 75-80

M.F. Ahmad, Associate professor in Coastal Engineering, head of department of Maritime Technology UMT, Malaysia. Co- researcher at some marine corrosion researches and co authors at several marine corrosion publication

Noor, C.W.M, Lecturer at University Malaysia Terengganu in subjects: Fluid Mechanics, Material Science. Completed research: Corrosion influence on mild steel welding seam.

A. Fitriadhy, Senior lecturer University Malysia Terengganu, Malaysia in subject: Advance Naval Architect. Co researcher at marine corrosion research: Influence of flow velocity and angle of attack on mildsteel coupons immerse in seawater (ongoing).

OWNERSHIP COST DRIVERS IN SHIP DESIGN & CONSTRUCTION

P. Sanders, Y. Guedj, BMT Design & Technology, Melbourne, Australia

SUMMARY

There is a perception that Naval ships cost too much to buy, specify, operate, maintain, sustain and dispose of. The reality or otherwise of this perception can only be determined through a structured cost analysis approach. Through the application of such techniques the stakeholders can begin to understand and address the key issues driving the costs. This paper considers ship design from the perspective of key through-life, cost-driving subsystems and by designing the ship to minimize the costs associated with these systems. At the simplest level, the ship is there to provide the infrastructure to support these systems. The proposed ship design process wraps the ship around the systems rather than fitting the systems into the hull.

1. INTRODUCTION

1.1 BACKGROUND

Naval ship design has traditionally been more of a creative process than a scientific process. The identification and selection of concepts has been primarily guided by experience, design codes, rules of thumb and the imagination of the design team. The concept phase has traditionally lasted as long as the time and money.

The motivation of this paper is to present a novel and structured approach to Naval ship design; one that acknowledges the significance of through life costs.

The Axiomatic Approach to Design is used as an implementation framework for a system based model.

The paper reviews the traditional approach and introduces a methodology that places much greater emphasis on a systematic design process. This modified process has a methodical review at each stage to provide an optimised design for ship systems. The basis for this optimisation is cost and is driven by well-informed decision. When convergence is achieved, a hull is then wrapped around the systems to create the platform.

1.2 IMPORTANCE OF THE EARLY STAGES OF DESIGN

The capital cost of a vessel is the summation of all the labour and material costs involved in the construction including any overhead costs. The final price of a vessel includes allowances for capital cost financing, inflation and shipyard profit.

The complexity and the costs associated with the detailed ship design of Naval ships coupled with the length of time between the preliminary design and completion are the leading factors that determine the flexibility of contracts between a navy and a shipyard. This inherent flexibility and lack of definition are the leading causes for the increasing number of design change orders and final price escalation costs.

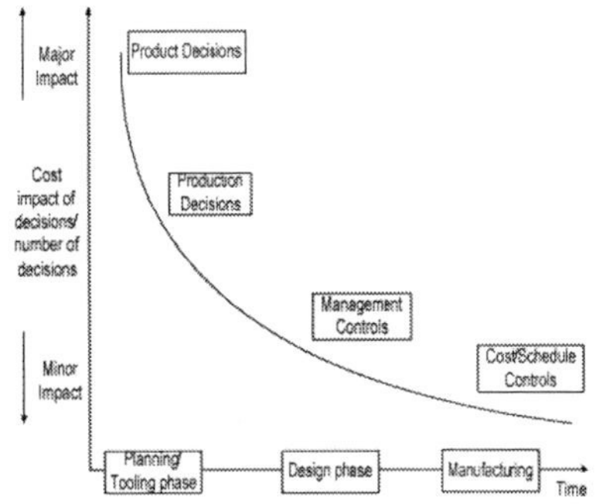

Figure 1 - Time Related Cost Impacts

Figure 1 gives a qualitative indication of the cost impact of decisions throughout a generic acquisition process. This relationship is well known, understood and is essentially intuitive. However, in practice many of the approaches taken in. ship design do not adhere to the principles and practices this relationship alludes to.

One can appreciate the major impact of decisions in the conceptual phase. Therefore, the ability to perform an analysis on performance/system versus cost trade-off in the early stage of the design process assists in understanding and optimising cost uncertainty during later design phase. The issue may not be one of design or process immaturity but rather current practices may be too mature and heavily reliant on techniques that are no longer appropriate (but hard to assess due to their subjective nature).

1.3 ACQUISITION COST VERSUS THROUGH LIFE COST

The drivers ruling the acquisition costs are well understood, and the highest level of through life costs are dominated by fuel and crewing, along with maintenance and repair. The challenge is therefore to minimise the through-life costs while providing a ship with the required operational capability. This trade-off is constrained by the need to demonstrate that any increase in acquisition costs will be offset by a through-life cost saving.

Recent work by BMT Design & Technology in Australia has shown that the true cost of ownership for Naval Ships can be reduced if due consideration is given during the design phase. This may be as simple as giving greater consideration to zonal design, removal routes, modular construction of key systems, considering the impacts of obsolescence of major systems or access to cabling and pipe work to facilitate maintenance and refit activities.

The views expressed in this paper are those of the authors and are not necessarily the official view of the BMT Group or any other organization. The intent of this paper is to foster dialogue to gain a better understanding of the ownership cost drivers of Naval ships.

2. TRADITIONAL DESIGN

2.1 TRADITIONAL SHIP DESIGN PROCESS

The Design Spiral: The traditional approach is relatively straightforward, well proven and effective in most cases but not necessarily efficient. The designer estimates the weight of the ship and every system to be included in the ship and balances this quantity with the underwater volume of the hull. The weight and underwater volume of the hull are distributed so that the operating ship floats upright and meets the performance requirements: speed and manoeuvrability.

Even though it seems simple in theory, the process is viewed as a highly coupled collection of interrelated physical attributes. For instance, the selection of the power of the ship propulsion system requires the knowledge of the resistance of the ship hull and the ship hull geometry cannot be entirely determined until the weight and volume to be carried is well defined. Evans' [1] general design diagram (shown in Figure 2) is a good representation of the process. The spiral spokes represent the set of all major areas that must be addressed through the design process to completely define the ship.

The spiral shape illustrates the current practice of independently developing each required parameter in a sequential manner, evaluating the relationship between design parameters, iterating to resolve conflicts and repeating the same process until all conflicts are resolved.

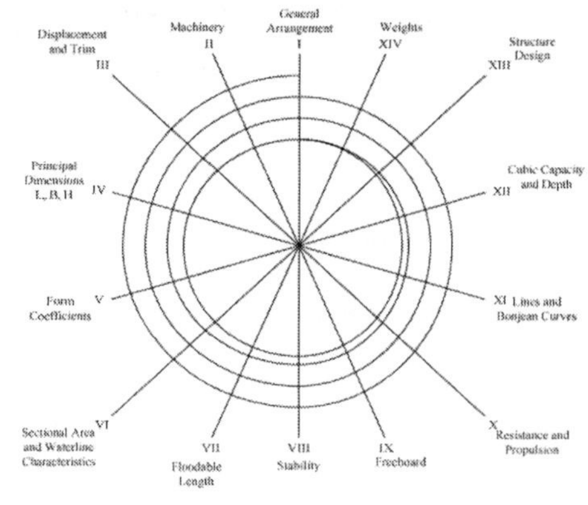

Figure 2- Evan's General Design Diagram [1]

The two principal limitations of the spiral are; that the process of design is assumed to be sequential and the opportunity to include life cycle consideration is limited.

Buxton [2] introduced economic issues into the spiral and time was added as a third dimension by Andrews. Andrews' representation is a helicoid and allows many dialogues and constraints to be shown to the designer as fundamental in the process (cf. Figure 3).

Despite the effort put into methods to expand the usefulness of the design spiral, the process stays iterative and so, laborious and expensive with little or no capacity to optimise for cost elements.

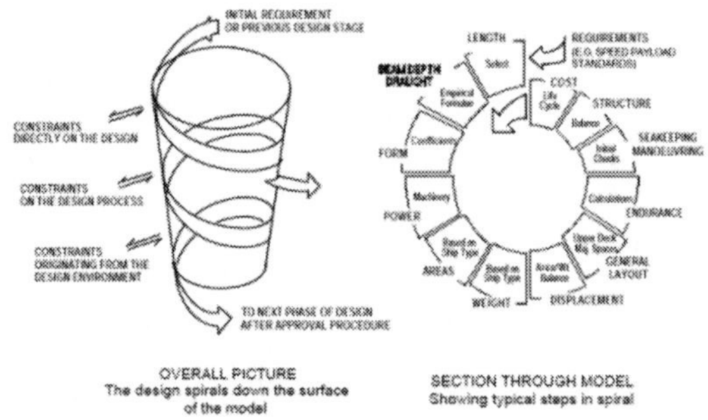

Figure 3 - Andrew's Overall of Ship Design Process

Lack of stability: The key point of the spiral model is the estimation of the design attributes, primarily weight and volume, in the early stages of the process. Therefore the more accurate the estimation, the quicker the design will reach convergence. In return, fluctuations and errors in the weight, volume and area estimates can have serious consequences on the design, which are hard to fix later on. This problem arises when the hull form is fixed too early in the design process, removing any flexibility from the design.

The hull is sized and shaped based on early estimates of the weight, volume and area, and the assumption that volume and areas are "arrangeable". The classic approach consists of virtually "fixing" an initial concept.

2.2 HISTORICAL HERITAGE

The traditional process is intuitive and has a long historical heritage. Early hull form design is motivated by the concern that hydrodynamic performance requirements need to be reached. But as the compromise between performance and space/weight requirement is a fundamental issue in hydrodynamics, the process might take time to converge into a stable design.

The hull is traditionally considered the primary system. Robert G. Keane mentioned the term "Outside- In" design approach where the design starts with the hull and then the hull is fitted with the ship's systems. [3]. He identifies the following impacts of the classic method:

- Non-optimum hydrodynamic hull form designs, which significantly increase energy consumption and the fuel bill for the fleet.
- Difficulty in maintaining and repairing the ships due to space limitations and the 'tightness' of the ship arrangement.
- Insufficient service-life allowances for weight and/or space, thus increasing modernisation costs.
- Significant reductions in terms of years of the economical service life of the ships.
- Possible operational restrictions due to the inability to develop robust designs where ship performance is minimally sensitive to variations in operations.

2.3 RISK REDUCTION

A current practice to reduce the risks associated with a new design is to pick an existing design that is known for its good performance and to add new systems in order to meet the requirements imposed by the customer. Although this method can reduce the costs associated with design effort, it appears that a large amount of effort is put into modifying the design in order to accommodate the new platform system.

3. ALTERNATIVE APPROACH

3.1 TOTAL SHIP SYSTEM ENGINEERING IN NAVAL COMBAT SHIPS

The term "Total ship system engineering" has recently been defined in the defence area to describe the fundamental system approach of modern naval engineering, and to provide a framework for ship designer's to follow.

Naval ship design is a far more complex process than the typical commercial ship design. The complexity of Navy ship design has long been acknowledged and the need to take a systems approach is well recognised if not widely adopted:

> "The complexity associated with the engineering of warship concepts is observed when considering the multiplicity of functions desired and the large number of physical subsystems and parts. The fact that the system must be considered becomes obvious if one considers that inserting a single highly advanced warship as a node into and existing battle group, that interoperability cannot be obtained since the equipment processing and interconnective protocols are incompatible". [4]

3.2 APPROACH TO KEY DRIVERS FOR DESIGN OPTIMISATION

As stated previously, traditional Naval ship design has been something of an art with the solution being crafted from the hand-on experienced industry professionals, mainly Naval Architects. In seeking to identify and explore the drivers for ownership costs it became clear that a more systematic whole ship design approach would be required. This would allow the various relationships to be identified, quantified and analysed.

A system engineering process approach was recognised as being essential in the synthesis of a solution, including:

- A library of ship design models that could be assembled to produce a toolbox of design options for later optimisation.
- A methodology for multi-objective decisions based on effectiveness, cost and risk. These factors cannot be easily combined as in commercial ships where the trade-off is simply against cost. For Naval ships the many objectives must be presented clearly and separately to facilitate trade-off and decision-making. This process is somewhat subjective.
- Practical and quantitative methods for measuring effectiveness and performance. A Measure of
- Effectiveness (MOE) and Measures of Performance (MOP) models are an essential requirements for optimisation and design trade-off.
- Qualitative and quantitative methods for measuring risk. Risks factors to be considered include performance, time and cost.
- A Life Cycle Cost Model sensitive to key characteristics whilst having the granularity appropriate to a concept design.
- A database of target ships with known characteristics and costs; this is to facilitate the exploration of concept options and to allow the various design modules to be populated with credible data.

- A 'trade-off' engine for refining concept development in a multi criteria and multidisciplinary design space.
- A way to visually represent the concept designs, in this case a 3D modelling tool was considered useful.

So, in simple terms the objective is to set up a series of models and decision matrices to allow a trade-off between ownership costs and effectiveness/performance. This concept is illustrated in Figure 4.

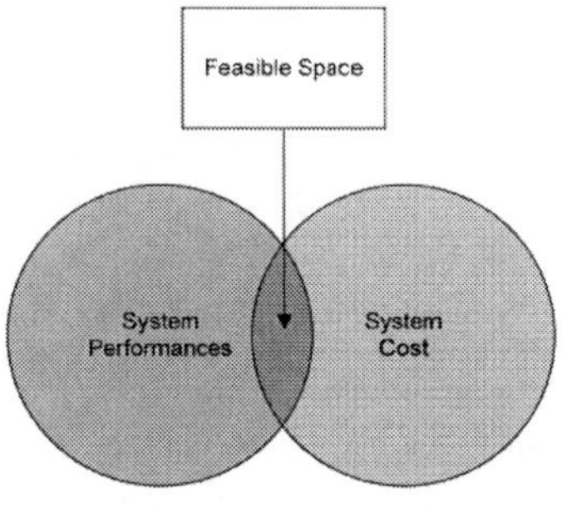

Figure 4 - Cost/Performance trade off

Having now established the essential components of the ownership cost model, a method to translate the customer needs into a model and ultimately develop a concept design is required.

3.3 AXIOMATIC APPROACH TO DESIGN

Axiomatic Design is a systems design methodology using matrix methods to systematically analyse the transformation of customer needs (CRs) into functional requirements (FRs), design parameters (DPs) and process variables (PVs). The methodology has been in development since 1990 by Dr. Suh Nam Pyo at MIT, Department of Mechanical Engineering [5]. The product development process is divided into four domains that are connected by three mappings, as shown in Figure 5.

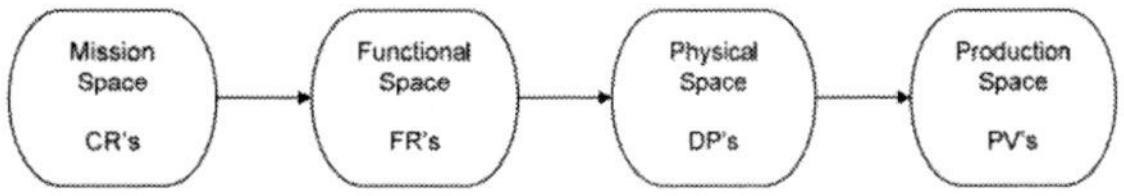

Figure 5 Design Mapping [5]

"The ultimate goal of the Axiomatic Design is to establish a science base for design, and to improve design activities by providing the designer with a theoretical foundation based on logical and rational thought processes and tools." [5]. While the classic and intuitive engineering design process is iterative, the Axiomatic Design process consists of generating basic and independent requirements, and focuses on selecting the best means to achieve the requirements. It breaks down a complex engineering problem into "simple" sub-problems, which are easier to solve:

- The *customer attributes/requirements* (CAs/CRs) represent the constraints imposed by the customer.
- The *functional requirements* (FRs) represent the goals of the design.
- The *design parameters* (DPs) express how the functional requirements will be satisfied.

The FRs and DPs can be described as vectors. The relationship between the FRs and the DPs can be stated as a matrix. This matrix is called the *design matrix*.

The design is defined as the *mapping process* from the functional space to the physical space to satisfy the designer-specified functional requirements.

The mapping process shall verify two fundamental axioms:

- The *Independence Axiom*: Maintain the independence of the functional requirements.
- The *Information Axiom*: Minimize the information content of the design.

In order to satisfy the Independence Axiom, the design matrix must be either a diagonal or triangular matrix. A diagonal design matrix represents an uncoupled design, while a triangular matrix represents a decoupled design. All other matrices represent coupled designs that require redesign or a different approach.

3.4 AXIOMATIC APPROACH APPLIED TO SHIP DESIGN [4]

Figure 6 summarises the mapping process applied to ship design after basic ship requirements have been defined (cf. Table 1).

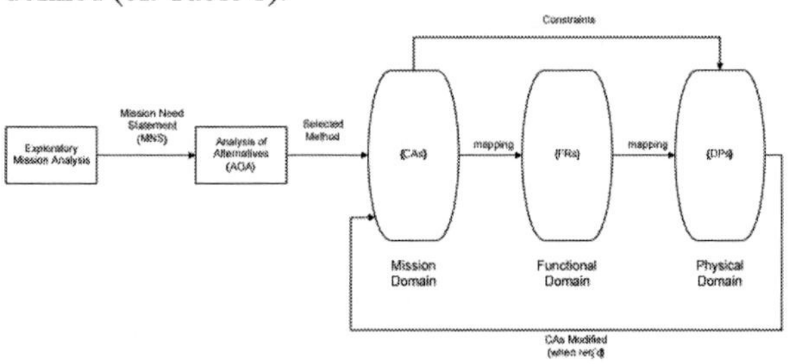

Figure 6 Axiomatic Approach Applied to Naval Engineering

Table 1 - High Level Customer Requirements

Customer Attributes (CAs)	
Maximum Speed	Vs
Operating Speed Range	Ve
Range of navigation	E
Stores Period	Ts
Manning	Nc

The application of the axiomatic approach to the early ship design stages consists of first identifying the *constraints* governing the design, then identifying the functional requirements and the associated design parameters.

The high level ship design constraints are summarised in Table 2. Two different sets of constraints can be identified: cost related constraints (C1, C2) and physics governing constraints (C3, C4..., C10).

Cost related constraints define a limit range for acquisition and life cycle costs as initial constraints imposes a cost oriented philosophy on the design. The following constraints arise from the concern of respecting a, "prudent naval architecture practice" as the heritage of an art performed and developed for hundreds of years.

Table 2 - Design Constraints of the Ship Design Process

Constraints	
C1	Initial acquisition cost < $XXM
C2	Average hourly operation cost < $XXM
C3	Full load displacement ($Disp_{FL}$) = Total Weight (W_t)
C4	Ensure upright stability, GM > 0 m
C5	Maintain acceptable transverse dynamic stability, 0.090 < GM/B < 0.122
C6	Installed propulsive power $_{Pi}$ > Required propulsive power P_{iREQ}
C7	Installed electrical power kW_G > Required electrical power kW_{GREQ}
C8	Total available volume > Total required volume
C9	Total available area > Total Required Area
C10	Incorporate design growth margins (weight, KG, propulsion and electrical power)

The six functional requirements (FRs) necessary for defining the high level function of a warship are formulated and listed in Table 3. The relevant design parameters (DPs) associated to each functional requirement are listed adjacent to their respective FRs.

Table 3 - Functional Requirements and Design Parameters of the Ship Design Process

Nb	Functional Requirements	Design Parameters
1	Move through water	Propulsion system
2	Maintain desired course	Manoeuvring and control system
3	Neutralize enemy targets	Combat system
4	Protect from enemy attack	Countermeasure methods
5	Conduct sustained underway operations	Support/Auxiliary systems
6	Operate on surface of water	Hull form

The practical application of the axiomatic approach is to map the functional domain to the physical, and to verify the Independence Axiom by formulating a decoupled design.

Xs and Os are used to represent the interaction between FRs and DPs.

Xs represent a strong functional dependence between FRs and DPs, xs represent a weak functional dependence, and Os represent independence.

Equation 1 is the representation of the design matrix mapping the ship's high level requirements to the ship's high level design parameters in its very basic form.

One can notice a matrix without peculiar property, which translates to a coupled design.

$$\left|\begin{matrix} FR1 \\ FR2 \\ FR3 \\ FR4 \\ FR5 \\ FR6 \end{matrix}\right| = \left|\begin{matrix} X & x & O & O & O & X \\ x & X & O & O & O & x \\ X & X & X & O & O & x \\ X & X & X & X & O & x \\ X & X & O & O & X & X \\ X & X & X & X & X & X \end{matrix}\right| \left|\begin{matrix} DP1 \\ DP2 \\ DP3 \\ DP4 \\ DP5 \\ DP6 \end{matrix}\right|$$

Equation 1 - Highest level design equations in their original forms

Whitcomb and Szatkowski [4] propose logical justifications for disregarding functional dependency observed in Equation 1 and introduce three additional constraints (cf. Table 4)

.

Table 4 - Additional Design Constraints

C11	Always operate at the design waterline
C12	Carry adequate fuel to transit endurance range E at endurance speed V_e
C13	Insure a sound structural integrity

Their analysis results in the formulation of a decoupled design, illustrated by a lower triangular design matrix (cf. Equation 2) where the Independence Axiom is respected and the Information Axiom is satisfied.

$$\left|\begin{matrix} FR1 \\ FR2 \\ FR3 \\ FR4 \\ FR5 \\ FR6 \end{matrix}\right| = \left|\begin{matrix} X & O & O & O & O & O \\ x & X & O & O & O & O \\ O & O & X & O & O & O \\ O & O & X & X & O & O \\ X & O & O & O & X & O \\ X & X & X & X & X & X \end{matrix}\right| \left|\begin{matrix} DP1 \\ DP2 \\ DP3 \\ DP4 \\ DP5 \\ DP6 \end{matrix}\right|$$

Equation 2 - Highest Level Design Equations Resulting in the Decoupled Design (Independence Axiom Verified)

The above operation allows the representation of a simplified and systematized design, where the influence of a design parameter on the functional requirements is well defined, and understood.

The resulting synthesis model process is shown in Figure 7.

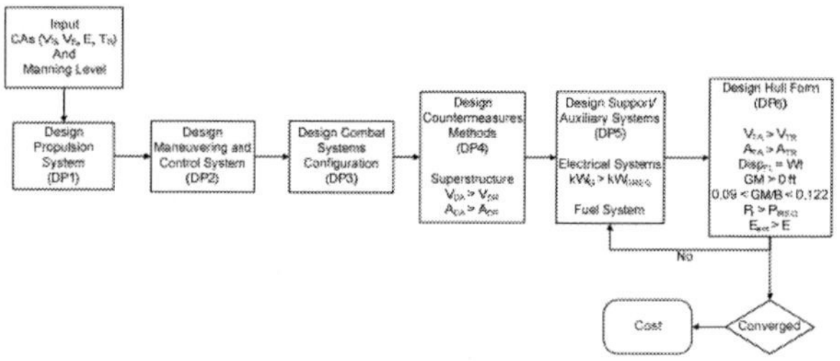

Figure 7 - Functional Ship Design Process (MIT)

The process starts with the design of ship's systems that define the weight, space and area requirements. The hull is finally designed using the quantities imposed by the design of the ship's systems (DP1, DP2, DP3, DP4, and DP 5) and the performances defined by the customer attributes (CAs).

The coherence of the design is assessed through a set of basic equations representing the design constraints.

3.5 IMPROVEMENTS TO SHIP DESIGN PROCESS AND NEW PERSPECTIVE

The axiomatic design theory, applied to the concept level ship design process, offers more control to the designer by eliminating the *ad hoc* assignment of design parameters found in the traditional design process. It also minimizes the need to modify functionally satisfactory design parameters once set. The approach allows a better understanding of systems interrelationships and integrates acquisition and through-life cost constraints in the development of a design solution.

More than a practical method, it is a mathematical formulation of a complexes process, an alternative conceptual representation of the traditional design spiral that can be used as the foundation for the development of new empirical design methodology.

The application of the Axiomatic Approach to Design to ship design demonstrates the importance of system definitions in the ship design process. This design philosophy opens a wide range of functional improvements to an inherently complex process. These improvements specially affect cost optimisation (cf. F) and decision making (cf. G).

3.6 A COST ORIENTED FUNCTIONAL SHIP DESIGN PROCESS

A cost concerned approach is achievable by integrating cost/performance trade-offs for each ship system into the functional ship design process (cf. Figure 8).

By this means, ship system costs are optimised during the design process and overall ship cost is assessed when the design has converged. This is a complex process for Naval ships.

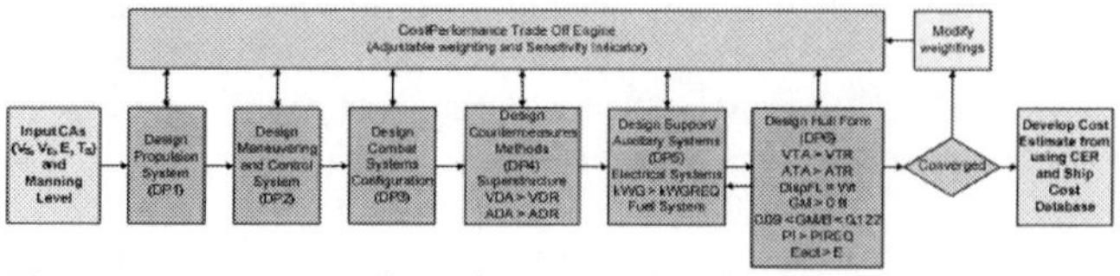

Figure 8 - Cost Oriented - Functional Ship Design Process (BMT D&T)

3.7 SET BASED DESIGN

A set-based design is applied along with the axiomatic approach for an effective ships system definition. The effectiveness of the design is achieved by the application of the following approach:

- Understand the design space;
- Integrate by intersection; and
- Establish feasibility before commitment.

For each system, the ranges of solution and associated requirements on other systems are communicated to each engineering discipline. Regions of feasibility are assessed by intersection of the different range of solution (cf. Figure 9). The range of solution for each discipline is reduced by elimination of subsystems solution that do not contribute to the total system solution. As the design space reduces in size, the level of detail increases and risk and cost estimates are evaluated with increasing accuracy.

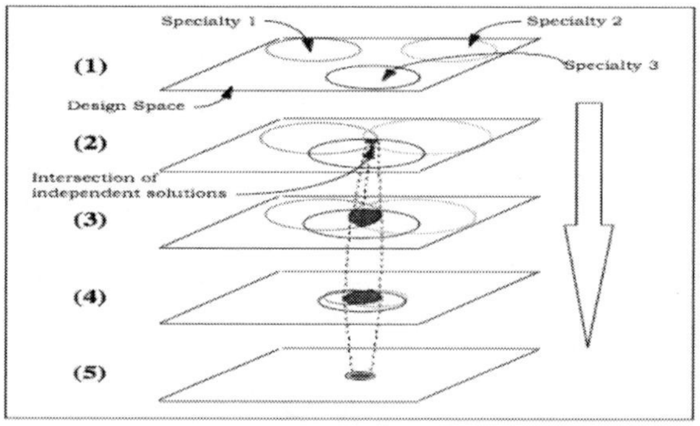

Figure 9 Set-Based Design process [10]

3.8 WRAPPING THE HULL AROUND THE SHIPS SYSTEMS

The axiomatic approach to design provides the means to clearly understand and outline the relationship between physical parameters and functional requirements. By doing this the designer can evaluate the impact of each successive design decision that occurs during the design process.

While traditional practices specify the design of the hull first, the axiomatic approach designs the platform systems first and then wraps the hull around the cumulative system volume and area.

This is completely contrary to the Design Spiral model and could therefore be met with opposition from conservative designers. Instead of trying to modify the initial concept to reach convergence and stability of design, the process focuses on designing the appropriate hull form to support the ship's systems. The ship is a platform that supports various complex systems and allows them to be moved through the water with performances fixed by the designer.

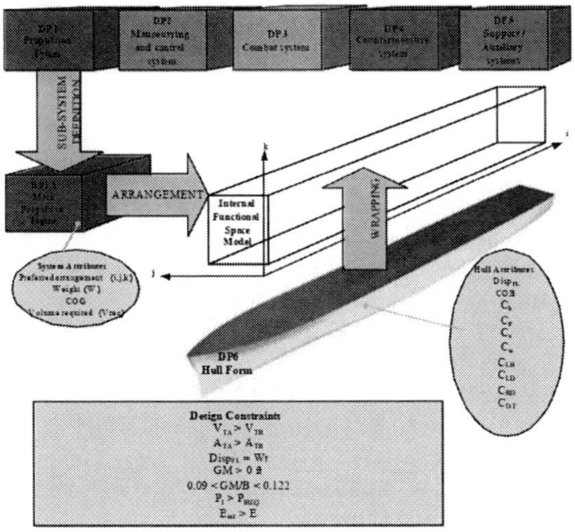

Figure 10 Design "Inside-Out"

The approach focuses on systems design and volume arrangement. Then the hull is wrapped around the systems while making sure that the constraints are respected and the performance is reached.

System arrangements: The arrangement operation described in Figure 10 consists in "packing" the internal functional space while respecting a defined set of rules. Each system is defined by a collection of attributes including: volume, weight and a set coordinates. The packing rules are defined as: first, all the ship systems are fitted into the internal functional space without overlapping; second, the global centre of gravity is centred in the internal functional space and the weight is correctly distributed; third, systems function and system inter-relationship are taken in account (e.g separation of accommodation and engines/machinery).

Arrangement (advanced packing) is then part of the well known class of mathematical problems that all relate to the allocation of a finite amount of supply to one or more items that consume part of this supply while respecting a set of rules.

There are a number of theories applicable for system arrangement. Fuzzy logic (cf. Figure 11), derived from fuzzy set theory, is used to aid placement of systems. The main principle of fuzzy logic is that it does not use the binary true or false, but a range of values. This means it can represent a relationship as a function, with a range of 0 to 1, indicating the best fit. Without fuzzy logic, an algorithm may find itself unable to place a system in its optimal position and end without producing a feasible design.

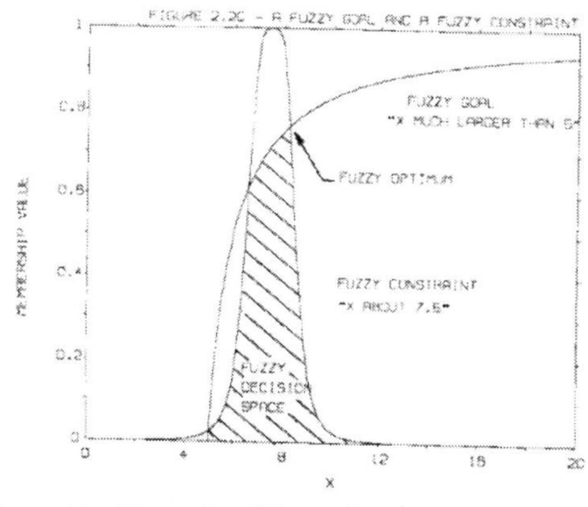

Figure 11 - Example of Fuzzy Logic

The theory allows for trade-offs between multiple criteria. For example, if accommodation is preferred to be away from machinery but this increases maintenance costs, a compromise would have to be reached. This could result in a tolerable noise level with less than optimal maintenance costs.

If every system was placed in the best spot for one condition, another factor would be impacted. The concept should take these relationships into account when generating a design. For example, the most satisfactory arrangement may include some systems that do not optimise their own maintenance costs, but have a better value for time to general quarters, thus increasing efficiency.

By applying weighting to criteria certain requirements can be optimised over others, this enables refinement of the design by altering the weighting of each criterion.

Hull Form Definition: After system arrangement, a template hull is fitted to the internal functional space. The hull is resized and wrapped over the internal functional space.

The hull form attributes comprise displacement, centre of buoyancy, hull size and hull form coefficients.

While the hull is modified to support the systems, hull attributes are benchmarked with data distribution obtained through regression analysis (cf. Figure 12). The balance of the solution produced is assessed through the set of constraints (Table 2 and 4).

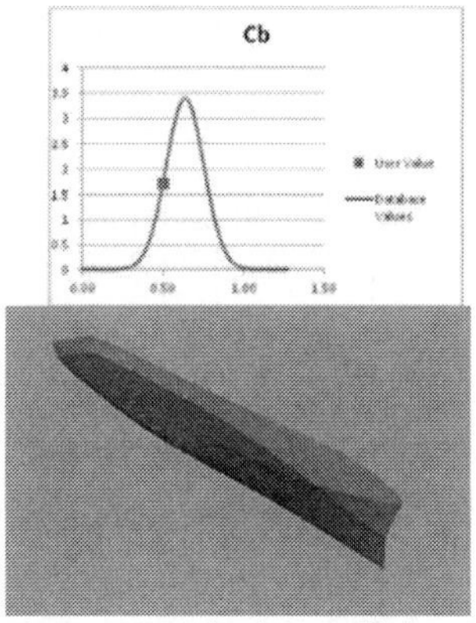

Figure 12 - Block Coefficient Definition

While compromises are made in both compartmentalization and hull performances, this approach results in an overall functionally efficient base design where design control and design flexibility are preserved until the achievement of the concept stage.

4. CONCLUSIONS

This paper has explored some of the issues and techniques around optimisation of a ship design to minimise total life-cycle costs.

The objective of this work is not to present an all-inclusive and comprehensive solution, instead the work demonstrates that optimising for total life cycle cost is a complicated multivariable problem that contains, and will always contain, a high degree of subjectivity even using systematic methods such as the Axiomatic Approach to Design.

There is merit in this approach (even at a high level) as it makes the owner, operator and designer aware of the through life implications of decisions made early in the acquisition process.

Ship design is a discipline with an inherent complexity commensurate to the ambition of a project. Finding methods to understand the design process, and therefore have a better control on the design itself, is a major objective for modern naval architecture.

Developing a common framework shared by naval architects, and system engineers, will lay the foundations required for a more functional design and a cost effective design.

The aim is to take this work forward by strengthening and validating the model developed and incorporation of greater granularity and certainty in the Cost Estimate Relationships that have been used. As well as cost, other aspects of Asset Management (such as Quality of Service and reconfigurability) will be added to the model.

The final output of the work will be a greater understanding of the issues and compromises related to optimisation of design from a total cost of ownership perspective. This understanding will be embedded into a decision-aid tool used to educate and inform those involved in the design, build, operation and maintenance of ships.

5. ACKNOWLEDGMENT

BMT Design & Technology wishes to acknowledge Qinetiq Consulting for the Paramarine Software that has been used in the development of the ideas contained in this concept paper.

6. REFERENCES

[1] Evans, J. Harvey, 1959, "Basic Design Concepts," *Naval Engineers Journal*, Vol. 21, Nov.

[2] Buxton, L. Ian, 1972, RINA publication.

[3] Keane R.G., "Reducing Total Ownership Cost : Designing Inside Out of the Hull", Ship Design USA, Inc.

[4] Whitcomb C.A., Szatkowski J.J, 2000 "Concept Level Surface Combatant Design in the Axiomatic Approach to Design Framework", First International Conference on Axiomatic Design, Cambridge, MA.

[5] Suh, N.P., 1990, *The Principles of Design*, Oxford University Press, New York.

[6] Mistree F., Smith W.F., Bras B.A., Allen J.K, Muster D., 1990, "Decision-Based Design: A Contemporary Paradigm for Ship Design".

[7] Bart van Oers and Hans Hopman, "A 3D Packing Approach for the Early Stage Configuration Design of Ships".

[8] Capt. Norbert Doerry, USN, "Using the Design Structure Matrix to Plan Complex Design Project".

[9] Brown, A.J., Thomas, M. 1998, "Reengineering the Naval Ship Concept Design Process", From Research to Reality in Ship Systems Engineering Symposium, ASNE.

[10] Bernstein, Joshua I. 1998, "Design Methods in the Aerospace Industry: Looking for Evidence of Set-Based Practices," Master of Science Thesis, Massachusetts Institute of Technology.

DEVELOPMENT OF CORROSION RESISTANT STEEL FOR BALLAST TANK

K. Shiotani, I. Samusawa, S. Tachibana and **T. Komori,** JFE Steel Corporation, Japan

SUMMARY

The superior corrosion resistance of a newly-developed corrosion resistant steel for ballast tank application was presented. First, the actual environment at the upper deck in a ballast tank of a bulk carrier was measured. The upper deck environment was characterized by cyclic repetition of water evaporation and condensation accompanied with daily temperature increases and decreases. The corrosion resistance of steels was evaluated by the cyclic corrosion test with salt spray using conditions found from the environmental measurement results. The degradation area of coating films at a scratch in the developed steel was about 70% that in the conventional steel. The fine rust particles which formed under the coating film on the developed steel were considered to contribute to its excellent corrosion resistance. The mechanical properties of the developed steel, such as tensile properties and Charpy impact properties, satisfied the KE36 specification.

1. INTRODUCTION

When a ship sails with no cargo, seawater is loaded in ballast tanks. Since the tanks are a severe corrosive environment, heavy-duty coatings are applied to prevent corrosion [1,2]. Strict specifications for coatings and their pretreatments are provided under the rules of the Performance Standard for Protective Coatings (PSPC) [2]. As the target life under PSPC is 15 years [2], recoating may possibly be needed during the typical ship design life of 25 years [3,4]. Since recoating is extremely expensive, development of corrosion resistant steel for ballast tank application is also necessary from the viewpoint of LCC (Life Cycle Cost) reduction.

The concept of development of a new corrosion resistant steel for extended coating life is shown in Fig. 1. In this paper, coating life is defined as the period until recoating. The degradation process is divided into two stages. The first stage is the period with no degradation of the coating film. The second stage is the corrosion stage. Corrosion starts at defects of coating films, such as pinholes, thinned coating regions, regions of weakened adhesion between the coating film and steel, and parts with mechanical damage. The expansion of degradation is related to the corrosion reaction of steels [5,6].

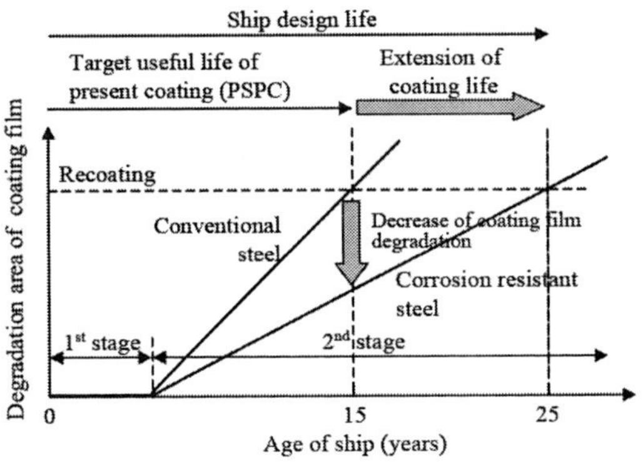

Figure 1: Concept of development of corrosion resistant steel for extended coating life.

The degradation rate can be reduced by application of corrosion resistant steels. The target of development of the corrosion resistant steel in this work is to extend the coating life from 15 to 25 years, corresponding to the typical design life of a ship.

In this paper, the corrosion environment at the upper deck in the ballast tank of an actual bulk carrier was measured. The corrosion resistance of steels was then evaluated by a cyclic corrosion test with salt spray using conditions found from the measurement results of the actual upper deck environment. The mechanical properties of the developed steel, such as the tensile properties and Charpy impact properties of plates and welds, were also evaluated.

2. EXPERIMENTAL

2.1.1 MEASUREMENT OF CORROSION ENVIRONMENT AT UPPER DECK IN ACTUAL BALLAST TANK

It is known that corrosion in the upper region of ballast tanks is intense compared with that in the lower region, and in particular, corrosion is heaviest near the upper deck. Therefore, corrosion environment measurements were made at the upper deck of a ballast tank of an actual bulk carrier. Fig. 2 shows the installation position of the environmental measurement system. The measurement system was installed in the No. 3 starboard side ballast tank (Fig. 2(a)). The system comprised a temperature sensor, humidity sensor, ACM (Atmospheric Corrosion Monitor) sensor, data logger, and battery. The degree of wetness can be understood by analyzing the ACM sensor output [7,8]. The measurement devices were installed near the upper deck about 2m from the hatch side girder (Fig. 2(b)). The temperature and humidity sensors were installed about 130mm below the upper deck (Fig. 2(c)). The ACM sensor was attached directly to the upper deck (Fig. 2(c)). The data logger and battery were stored in a waterproof box. Temperature and relative humidity (RH) data and the ACM sensor output were recorded every 10 minutes. The appearance of the installed devices is shown in Fig. 3. The voyage route during the measurement was a round-trip from Japan to Australia.

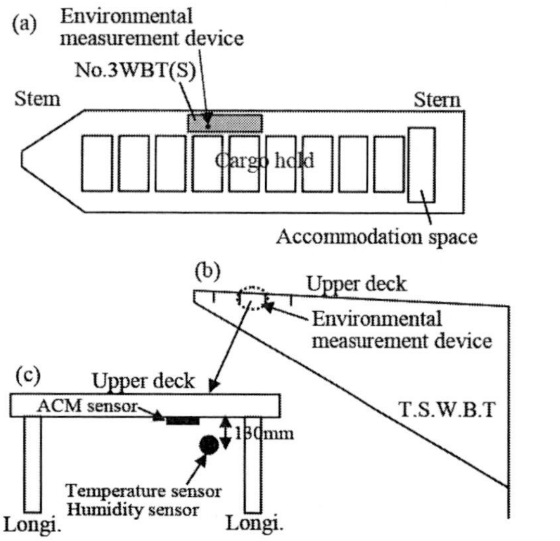

Figure 2: Installation position of corrosion environment measurement system.
(a) Ballast tank in which device was installed
(b) Position of device in ballast tank
(c) Positions of temperature sensor, humidity sensor, and ACM sensor

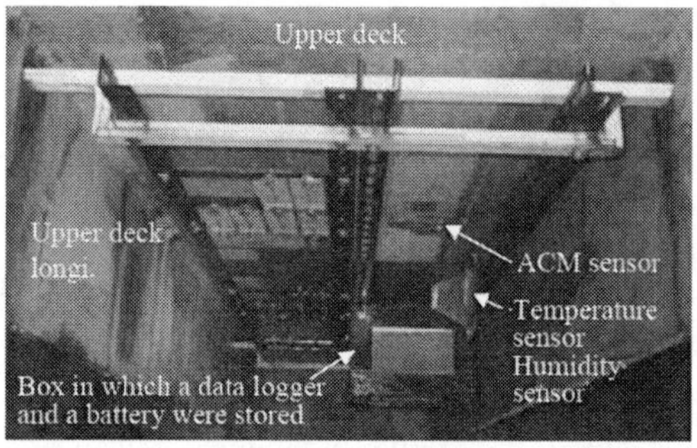

Figure 3: Appearance of installed environmental measurement device.

Seawater was loaded in the ballast tanks on the way from Japan to Australia. No seawater was loaded in the tanks from Australia to Japan. The seawater height in the ballast tank from Japan to Australia almost reached the upper deck. The measurement was performed from Feb. 5 to Mar. 28, 2008.

2.1.2 EVALUATION OF CORROSION RESISTANCE OF STEELS

The corrosion resistance of the developed corrosion resistant steel and a conventional steel for comparison was evaluated by the cyclic corrosion test with salt spray using conditions found from the environmental measurement results at the upper deck in the actual ballast tank, as described later.

Three specimens with dimensions of 3mm in thickness, 150mm in width, and 300mm in length were taken from both steels.

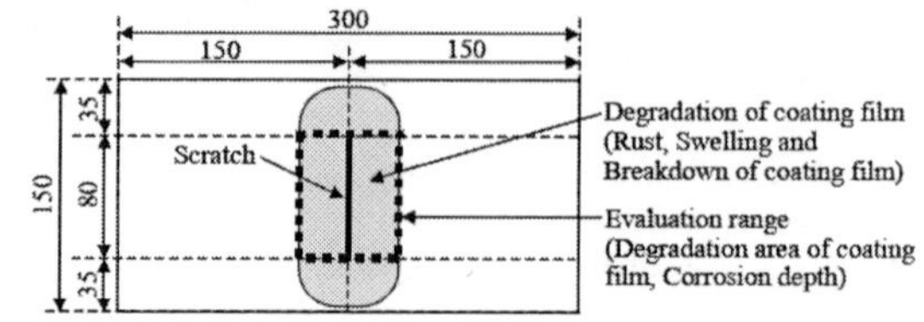

Figure 4: Specimen size, position of scratch, and evaluation range (degradation area of coating film, corrosion depth).

After blasting the specimen surfaces, the specimens were coated with an inorganic zinc primer with a thickness of 19μm, followed by coating with a modified epoxy. Epoxy coating was performed two times with a target coating thickness of 160μm each time. Therefore, the actual total thickness of the epoxy was about 350μm. A scratch 80mm in length, which reached the steel surface, was made with a retractable knife, as shown schematically in Fig. 4, in order to evaluate the degradation of the coating films in the second stage in Fig. 1.

Next, the corrosion resistance of the specimens was evaluated by the cyclic corrosion test. The test cycle was as follows : (1) 35°C, 5%NaCl spray, 2h → (2) 60°C, 20-30%RH, 4h → (3) 50°C, >95%RH, 2h (return to (1)). This cyclic test was carried out for 772 days, over 2300 cycles in total. During the test, the appearance of the specimens was photographed periodically, and the degradation areas of the coating films around the scratch were measured from the photos (Fig. 4). The degradation area was defined as the sum of the rust area (breakdown area of coating film) and the swelling area of the coating film. After completion of the test, the remaining coating films and rust were removed from the specimens, and the corrosion area and corrosion depth were measured. The measured values were converted to values per 10mm length of scratch.

2.1.3 MECHANICAL PROPERTIES OF STEELS

The tensile test and Charpy impact test of plates and welds of the developed corrosion resistant steel were performed in accordance with the Rules for the Survey and Construction of Steel Ships (Part K: Materials [9] and Part M: Welding [10]). Plates 33mm in thickness, which were manufactured by the actual production mill, were used for mechanical testing. U1 tensile test specimens, with a gauge length of 200mm, were taken from the transverse direction (C-direction) of the plate, and U4 impact test specimens were taken from the longitudinal direction (L-direction). Tensile and Charpy impact tests of the plate were performed at the ambient temperature and −40°C, respectively.

For welds, FCB (Flux Copper Backing) welding with a heat input of 214 kJ/cm was performed, and U2A tensile test specimens and U4 impact test specimens were taken from the welds. The impact test specimens were taken from 1mm below the surface, and the notch positions were the center of weld metal, the bond, and 1, 3 and 5 mm from the fusion line. The test temperatures were the ambient temperature for the tensile tests and –20°C for the impact tests.

3. RESULTS AND DISCUSSION

3.1.1 CORROSION ENVIRONMENT AT UPPER DECK IN ACTUAL BALLAST TANK

The environmental measurement results at the upper deck in the actual ballast tank are shown in Fig. 5. At first, the average temperature was about 8°C while the ship was in Japan. After leaving Japan, the temperature rose, reaching 25 to 50°C on the voyage from Japan to Australia, in Australia, and while returning from Australia to Japan. The temperature gradually fell as the ship approached Japan.

The maximum temperature during.the voyage was very high, reaching as much as 53°C. Temperature, relative humidity (RH) and the ACM sensor output varied cyclically as shown in Fig. 5. Fig. 6 shows details of the results from Feb. 23 to Feb. 28. The temperature was about 25-30°C from 0:00 midnight to 7:00 a.m., rose to about 40-50°C at around 2:00p.m., and fell to 25-30°C at midnight. An inverse trend of RH values and temperatures is noted. The RH values were in the range from 40 to 70% at the highest temperature on each day (around 2:00p.m.) and were almost 100% at 7:00a.m. This falling and rising RH is due to the rising and falling temperature during the day. The ACM sensor output decreased as the RH values decreased, and it increased as the RH values increased, suggesting that water evaporated at low RH values and condensed at high RH values.

Figure 5: Environmental measurement results at upper deck in actual ballast tank.

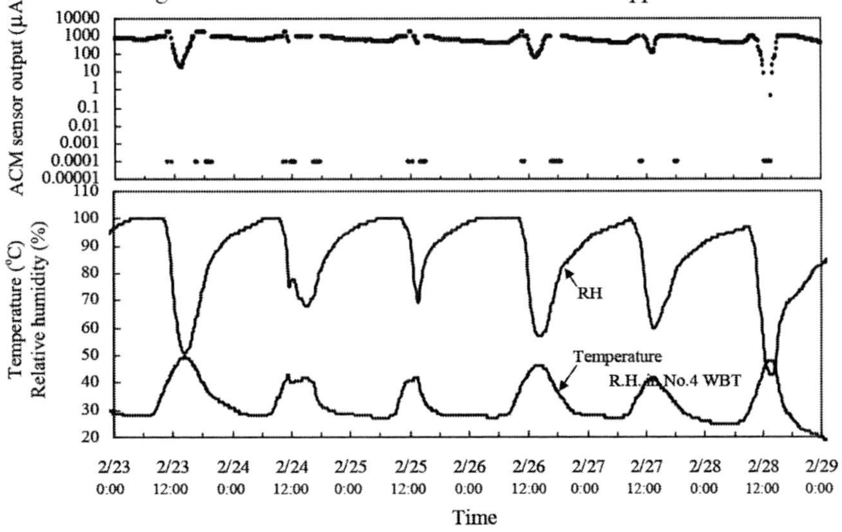

Figure 6: Details of measurement results of temperature and relative humidity and ACM sensor output.

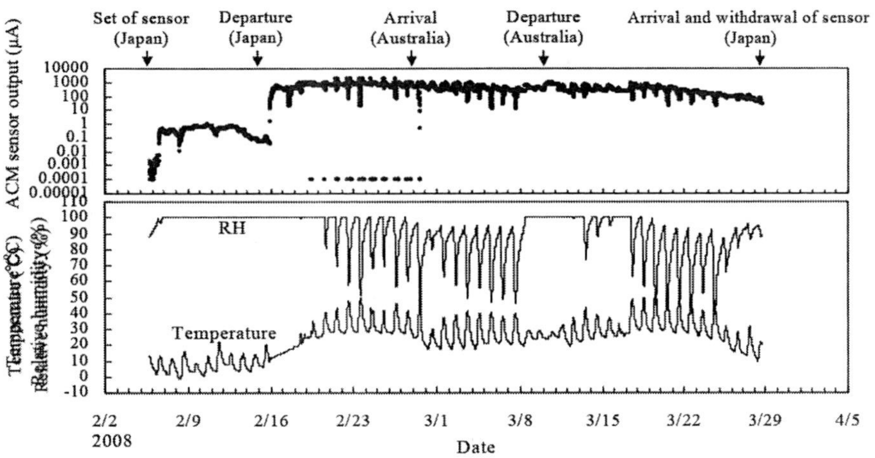

The cyclic output of the ACM sensor indicates that the ACM sensor is not immersed in seawater. Regarding adhesion of sea salt to the ACM sensor, it has been reported that ACM sensor output was about 20 μA with sea salt adhesion of 100 g/m^2 at RH of 50% [8]. In the data at around 2:00p.m. on Feb. 23 in Fig. 6, the ACM sensor output was about 20μA at RH of about 50%. This suggests that the amount of sea salt adhesion to the ACM sensor was also about 100 g/m^2 in this investigation.

Since the corrosion rate changes depending on the thickness of the water film on steels [11-13], the thickness of the water film is one of the most important factors in the corrosion reaction of steels. It has been reported that the corrosion rate with thin water films with thicknesses in the range of about 20-110 μm is much larger than that with thick water films such as immersion condition [13], due to more supply of oxygen to steel surface. Thickness of water film can be estimated by the amount of deposited sea salt and RH [13]. In case of the deposited sea salt of 100 g/m^2, which is estimated at the upper deck in this study, RH for the formation of the water film of 110 μm or less in thickness is estimated to be 73% or less. The condition with RH of 73% or less, which caused higher corrosion rate than that of immersion condition, occurred 24 times in 52 days during this voyage in Fig. 5, and its total time was 156 hours. It is one of causes for severe corrosion of upper decks that very thin water film forms in high frequency by cyclic repetition of water evaporation and condensation.

Fig. 7 summarizes the corrosion environment at the upper deck in the ballast tank. The key points regarding this environment are: (1) Exposure to high temperatures of around 50°C, (2) not immersed in seawater even when seawater is loaded, (3) sea salt adhesion, and (4) cyclic repetition of water evaporation and condensation processes accompanied with diurnal temperature changes.

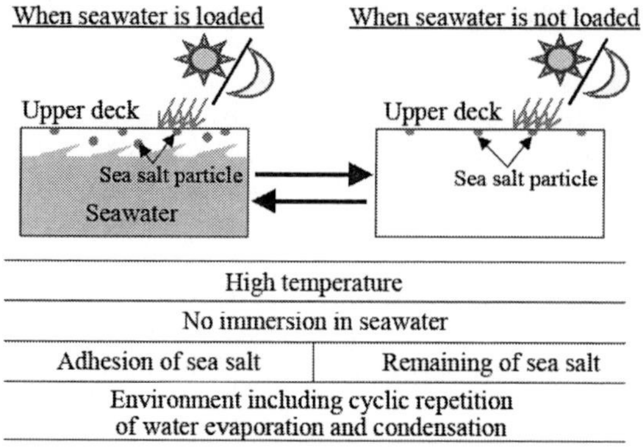

Figure 7: Corrosion environment at upper deck in ballast tank.

These factors should be taken into account in the test conditions to evaluate corrosion resistance in the upper deck environment in ballast tanks.

3.1.2 CORROSION RESISTANCE OF DEVELOPED STEEL

As described in section 2.2, the corrosion test conditions in this study were determined based on the actual environmental measurement results, and were as follows: (1) 35°C, 5%NaCl spray, 2h → (2) 60°C, 20-30%RH, 4h → (3) 50°C, >95%RH, 2h (return to (1)). This test includes all the factors of high temperature, no immersion, sea salt adhesion, and cyclic repetition of water evaporation and condensation, which are the key points in connection with the upper deck environment. This test is an accelerated corrosion test by high frequency repetition of the water evaporation and condensation processes. The appearances of the conventional steel sample and the developed steel after testing for 772 days are shown in Fig. 8.

Figure 8: Appearances of conventional and developed corrosion resistant steels after 772 days in cyclic corrosion test with salt spray. C2 and S1 show the ID No. of each specimen.

Steel	After corrosion test	After removal of coating film and rust
Conventional steel (C2)	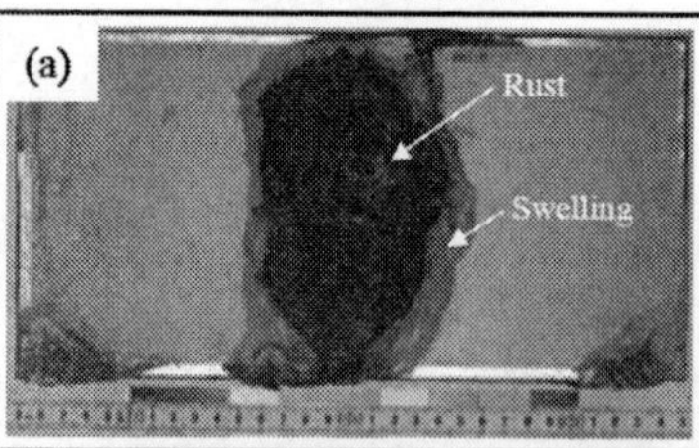	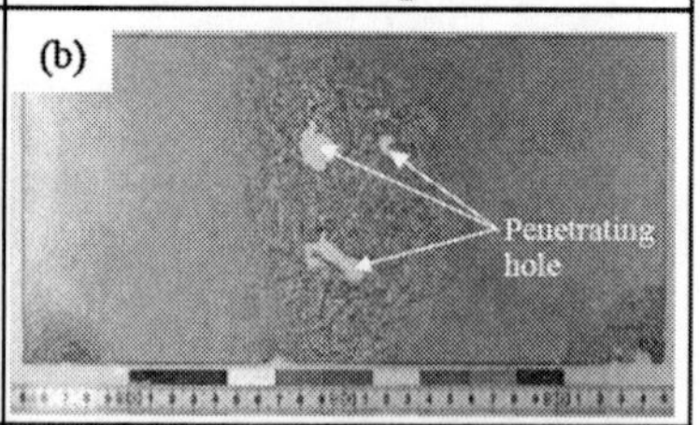
Corrosion resistant steel (S1)	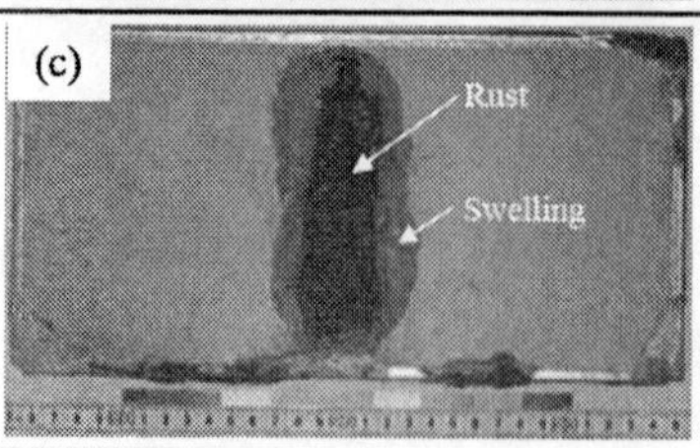	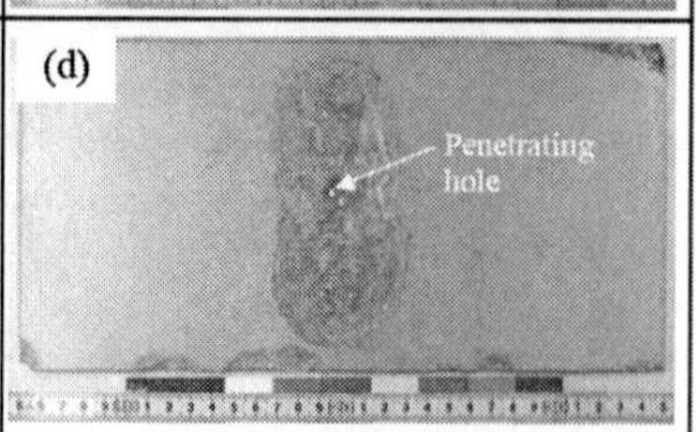

The left side of the figure (Figs. 8(a) and 8(c)) shows the appearance after the test, and the right side (Figs. 8(b) and 8(d)) show the appearance after removal of the coating film and rust. The degradation area of the coating film in the developed steel was clearly smaller than that in the conventional steel (Figs. 8(a) and 8(c)). Fig. 9 compares the progress of degradation of the coating films during the test in the conventional and developed steels. The degradation rate in the developed steel was smaller than that in the conventional one. The average degradation area of the developed steel was about 70% that of the conventional steel after 772 days of testing. For example, the time to reach a degradation area of 600 mm^2 with the conventional steel was about 500 days, while that with the developed steel was about 850 days by extrapolating plots. This means the period to recoating in the second stage in Fig. 1 is 1.7 times longer with the developed steel than with the conventional steel.

Nippon Kaiji Kyokai investigated the relationship between the condition of the coating of ballast tanks and ship age for a 2-coat tar epoxy coating with a 250μm film thickness [1]. According to the results, corrosion starts at about 5 years, and the coating condition changes from Good to Fair, as defined by the IACS Recommendation [14], at about 17 years in the best case. Recoating is recommended when the condition becomes Fair [14]. Since the PSPC coating specification [2] is superior to the above coating, it can be assumed that the progress of degradation in PSPC coating is similar to the best case for the above 250μm tar epoxy coating.

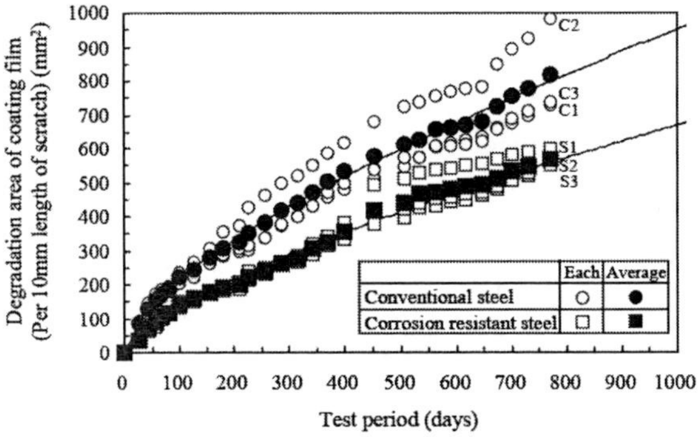

Figure 9: Transition of degradation area of coating films on conventional and developed corrosion resistant steels in cyclic corrosion test with salt spray. C1~C3 and S1~S3 show the ID No. of each specimen.

In this case, the period of the first stage in Fig. 1 is assumed to be 5 years, and the period to recoating in the second stage is assumed to be 12 years. With the developed steel, the period to recoating in the second stage can be calculated as 20 years or more, since the period to recoating in the second stage is 1.7 times longer than that of the conventional steel, as above described. Including the 5 years of the first stage, the time to recoating is calculated as 25 years or more. Therefore, it is expected that the time from initial coating to recoating with the developed steel will be 25 years or more.

The volume loss property of the developed steel is discussed below. Penetrating holes were observed in the samples after removal of the coating films and rust, as shown in Figs. 8(b) and 8(d). Since the sample thickness was 3mm, corrosion of the steels exceeded 3mm at the holes. As clearly shown in the figure, the total area of the holes in the developed steel was much smaller than that in the conventional steel. The measurement results of the corrosion area, corrosion depth, and volume loss are shown in Table 1. The average corrosion depth means the average depth in the corrosion region except for penetrating holes. The volume loss was approximated by the simple equation given in Table 1. The volume loss in the developed steel was about 60% that in the conventional steel. Therefore, prolongation of the time to replacement of plates due to corrosion can be expected by applying the developed corrosion resistant steel.

3.1.3 MECHANICAL PROPERTIES OF DEVELOPED STEEL

The mechanical properties of plates and welds of the developed corrosion resistant steel are shown in Table 2. The tensile and Charpy impact properties of the plates and welds satisfied the KE36 specification of Class NK [9-10].

Fig. 10 shows examples of transmission electron micrographs of rust formed on the conventional and developed steels. The fine rust particle which formed on the developed steel appear to contribute to its excellent corrosion resistance. However, further study is needed in order to clarify the precise mechanism of improved corrosion resistance, for example, characterization of the chloride ions which reach the steel surface.

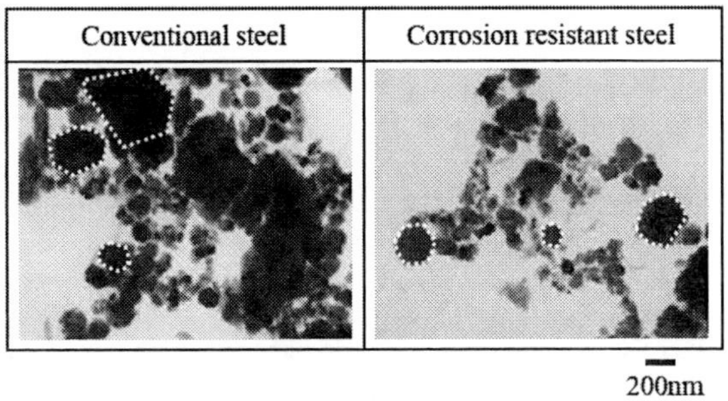

Figure 10: Transmission electron microscope images of rust particles formed on conventional and developed corrosion resistant steels under coating films after cyclic corrosion test with salt spray. Rust scraped from rusted steels was observed. The parts enclosed by the white dotted lines show one rust particle.

Table 1: Measurement results of corrosion area, corrosion depth, and volume loss.

	Conventional steel				Corrosion resistant steel				Ratio (%)
	Each			Ave.	Each			Ave.	(H)
	C1	C2	C3	(F)	S1	S2	S3	(G)	
(A)Total corrosion area*1) (mm^2)	712	965	723	800	592	557	544	564	71
(B)Corrosion area of 3mm or more depth*1) (mm^2)	126	73	147	115	2	39	27	22	19
(C)Corrosion area of less than 3mm depth*1) (mm^2)	586	892	576	685	590	518	517	542	79
(D)Average corrosion depth in part of less than 3mm depth (mm)	1.18	1.40	0.94	1.17	1.26	1.13	1.11	1.17	100
(E) Volume loss*1), *2) (mm^3)	1066	1467	980	1171	747	700	656	701	60

*1) Per 10mm length of scratch, *2) $E=3\times B+D\times C$, *3) $H=G\times 100/F$

Table 2: Mechanical properties of plate and welded joint of developed corrosion resistant steel.
(a) Plate, (b) Welded joint

(a)

Steel Grade	Tensile property			Charpy impact Property
	YS (MPa)	TS (MPa)	El (%)	Energy (J) (–40 °C)
Corrosion resistant steel (Thickness:33mm)	464	541	25	340
KE36 Spec.	≥355	490~620	≥20	≥34

(b)

Steel Welding condition Grade	Tensile property	Charpy impact Property		
	TS (MPa)	Notch position		Energy (J) (–20 °C)
Corrosion resistant steel (Thickness:33mm) FCB welding method (Heat input:214kJ/cm)	534	Face	WM	93
			BOND	116
			HAZ1mm	102
			HAZ3mm	143
			HAZ5mm	292
		Root	WM	131
KE36 Spec.	≥490			≥34

4. CONCLUSIONS

In this paper, the actual corrosion environment at the upper deck in the ballast tank was investigated. The corrosion resistance of the developed steel was examined in the cyclic corrosion test with salt spray based on the environmental measurement results. The mechanical properties of the developed steel plates and welds were also evaluated. The following conclusions were obtained.

(1) The actual environment at the upper deck in the ballast tank is characterized by cyclic repetition of water evaporation and condensation accompanied with daily temperature increases and decreases. The maximum temperature was very high, reaching 53°C. The upper deck was not immersed in seawater even when seawater was loaded. However, based on the ACM output, sea salt appeared to adhere to the upper deck to some extent.

(2) The degradation area of the coating film on the developed steel sample was about 70% that on the conventional steel in the cyclic corrosion test with salt spray for 772 days. Furthermore, the time to recoating with the developed corrosion resistant steel is predicted to be 25 years or more by considering the increase tendency of the degradation area of the coating film.

(3) The volume loss of the corrosion resistant steel sample was about 60% that of the conventional steel in the cyclic corrosion test with salt spray for 772 days. Therefore, prolongation of the time to replacement of plates due to corrosion can be expected by applying the developed corrosion resistant steel.

(4) The tensile and Charpy impact properties of plates and welds of the developed corrosion resistant steel satisfied the KE36 specification of Class NK.

ACKNOWLEDGEMENTS

The authors are grateful to KUMIAI SENPAKU CO., LTD. and Mitsui O.S.K. Lines, Ltd. for their cooperation in the environmental measurement of the actual ballast tank. Fruitful discussions with Dr. S. Mitao at Steel Research Laboratory, JFE Steel Corporation are also gratefully acknowledged.

REFERENCES

1. NIPPON KAIJI KYOKAI, 'Guidance for Corrosion Protection System of Hull Structures-For Water Ballast tanks and Cargo Oil Tanks- Second Revision', 1995.
2. Nippon Kaiji Kyokai, 'Guidelines for Performance Standard for Protective Coatings contained in IMO Resolution MSC.215(82) ', 2007.
3. IACS, 'Common Structural Rules for Double Hull Oil Tankers', 2006.
4. IACS, 'Common Structural Rules for Bulk Carriers', 2006.
5. W. FUNKE, 'BLISTERING OF PAINT FILMS AND FILIFORM CORROSION', Progress in Organic Coatings, 1981.
6. P. A. Sørensen, S. Kiil, K. Dam-Johansen, C. E. Weinell, 'Anticorrosive coatings: a review', J. Coat. Technol. Res., 2009.
7. S. Motoda, Y. Suzuki, T. Shinohara, Y. Kojima, S. Tsujikawa, W. Oshikawa, S. Itomura, T. Fukushima, S. Izumo, 'ACM (Atmospheric Corrosion Monitor) Type Corrosion Sensor to Evaluate Corrosivity of Marine Atmosphere', Zairyo-to-Kankyo, 1994.
8. T. Shinohara, S. Motoda, W. Oshikawa, 'Evaluation of Corrosivity of Atmosphere by ACM Type Corrosion Sensor', Zairyo-to-Kankyo, 2005.
9. CLASS NK, 'Rules for the Survey and Construction of Steel Ships (Part K MATERIALS)', 2005.
10. CLASS NK, 'Rules for the Survey and Construction of Steel Ships (Part M WELDING) ', 2005.
11. N.D.TOMASHOV, 'Development of the Electrochemical Theory of Metallic Corrosion', Corrosion, (1964)
12. M. YAMASHITA, H. NAGANO, R. A. ORIANI, 'DEPENDENCE OF CORROSION POTENTIAL AND CORROSION RATE OF A LOW-ALLOY STEEL UPON DEPTH OF AQUEOUS SOLUTION', Corrosion Science, 1998.
13. Y. Hosoya, T. Shionohara, W. Oshikawa, S. Motoda, 'Relationship between Corrosion Rate of Carbon Steel and Thickness of Water Film Containing Sea Salt', Zairyo-to-Kankyo, 2005.
14. IACS, 'Recommendation 87 GUIDELINES FOR COATING MAINTENANCE & REPAIRS FOR BALLAST TANKS AND COMBINED CARGO/BALLAST TANKS ON OIL TANKERS, Rev.1', 2006.

AUTHORS' BIOGRAPHIES

Kazuhiko Shiotani holds the current position of Senior Researcher at the Corrosion Resistant Steel Research Dept. of the Steel Research Laboratory, JFE Steel Corporation. He is responsible for developing corrosion resistant steels.

Itaru Samusawa holds the current position of Researcher at the Corrosion Resistant Steel Research Dept. of the Steel Research Laboratory, JFE Steel Corporation. He is responsible for developing corrosion resistant steels.

Shunichi Tachibana holds the current position of Senior Researcher at the Corrosion Resistant Steel Research Dept. of the Steel Research Laboratory, JFE Steel Corporation. He is responsible for developing corrosion resistant steels.

Tsutomu Komori holds the current position of Senior Researcher at the Corrosion Resistant Steel Research Dept. of the Steel Research Laboratory, JFE Steel Corporation. He is responsible for developing corrosion resistant steels.

EFFICIENTLY AND LESS CARBON EMMISION HYBRID CATAMARAN FOR SHORT DISTANCE ROUTES.

Ferry, M., WB.Wan Nik and MFAhmad, Universiti Malaysia Terengganu, Malaysia
Pelupessy, D and WME. Wattimena, Pattimura University, Ambon, Indonesia

SUMMARY

The increasing of fossil's fuel prices and environmental awareness urges engineers try to hybrid diesel with green energy such as solar energy. Through photovoltaic cells, solar energy is converted into electricity that could power the electric motors as the prime mover at a catamaran that has more slimmer and shallower cross-section area than single hull boat of an equivalent displacement has less draught, more deck area and more stable.. Electric motor maintains high efficiency over a wide range of loads and speeds has a small weight and need low maintenance and repair cost compared to diesel engines or outboard motors. At countries like Malaysia and Indonesia where solar energy is available during average six to seven hours a day is a quite good time range to charge the batteries until reach its full capacity, but when solar radiation is not available the battery charging is taken over by diesel generator. Such hybrid system is efficiently to be used for catamaran boat on rivers, between ports at coastal area and between closed islands due to its less the operational cost, maintenance cost and less carbon emission compare with fully diesel system, hence this is a more environmental friendly system.

NOMENCLATURE

C_b	Block coefficient
C_m	Midship coefficient
C_p	Prismatic coefficient
C_{wp}	Waterplane coefficient
C_T	Total resistance coefficient
EF_{CO2}	Emission factor (kg CO_2/fuel units)
GM_t	Transverse metacenter height (m)
KB	Height of center of buoyancy (m)
KG	Height of center of gravity (m)
L_{WL}	Length of waterline (m)
S	Wetted area (m^2)
SFOC	Specific fuel oil consumption (g/kWh)
R_T	Total resistance (kN)
V	Ship velocity (m/sec)
WSA	Wetted surface area (m^2)
ρ	Density of seawater (t m^{-3})

1. INTRODUCTION

A typical scenario for this study is based on sea activities in Malaysia and Indonesia, especially at regions which consists of islands and seas. Sea transportation between the islands within the region nowadays are mostly dominated by small monohull passenger boats powered by inboard diesel or outboard gasoline engines.

These outboard or inboard diesel engines during operation generate exhaust gasses which cause immeasurable negative effects on the environment, besides that the engine cooling water and waste water contains oil and fuel residue that is pumped out into the sea and cause pollution especially when there is a large number of boats operating in the region.

Moreover, due to the economic growth of these regions the number of boats will increase rapidly and consuming more fossil fuel and therefore causing more pollution. Progressively switching from fossil fuel to hybrid with alternative energy sources could reduce pollution and operational cost. Tropical region such as in Indonesia and Malaysia having an averrage 6-7 hours free solar energy that can be utilized to achieve this transformation [1], that is a hybrid solar- gasoline/diesel generator system which may initiate interest to boat operators due to the benefits they provide.

2. BOAT DIMENSION, WATER RESISTANCE AND POWER

2.1 MONOHULL BOAT

The design of a boat typically begins with analysis of the existing boats according to all of the requirements which are important to obtain general information on the type of particular boat that is the passenger boat. If the requirements will be fulfilled with the successful the design might proceed using this boat as the basis boat and, thus, involve scaling its characteristics for changes intended in the design. In this study, it is necessary bringing the design calculations and analysis of monohull and catamaran which then fulfilled the requirement as a passengers boat [2]. Also evaluate the performance including the hydrostatic, speed, resistance, and power estimation by maxsurf software.

Small passenger boats connecting closed islands most are monohull boats that carry 20-25 seated passengers, running at economical speed 10 knots, operate 300 days a year, 12 trips daily during 12 hours from 6.00 am – 6.00 pm is shown at Fig. 1. The body plan view of the monohull boat is shown at Fig. 2 and the main dimensions is determined by maxsurf software presented in Table 1. Based on the main dimensions is determined the deck space for passengers which has length 7m and breadth 1.85m, thus the deck for passengers is $13m^2$, excluded engine casing area the space for one passenger is obtained about $0.5m^2$.

Figure 1: Small diesel monohull boat

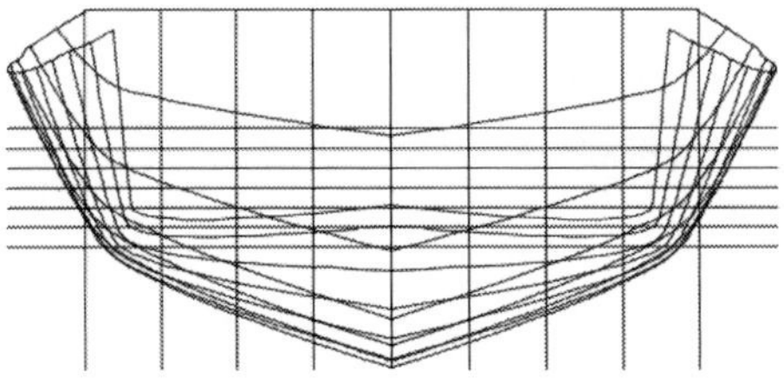

Figure 2: Body plan view of monohull boat

2.1 (a) Water Resistance and Power

Water resistance of a boat could be determined by fomula (1):

$$R_T = \frac{1}{2} C_T * \rho * V^2 * S, \quad (1)$$

Where : R_T – Total Resistance (kN)
C_T – Total resistance coefficient
ρ – Seawater density (t/m^3)
V – Boat speed (m/sec)
S – Wetted surface area (m^2)

2.1 (b) Diesel Power System

On a diesel engine propulsion system the rpm of the main engine is reduced by the gearbox with certain ratio so that the propeller rpm is reduced. The main engine speed is controlled by the boat driver through the engine speed controller, see Fig. 3.

Table 2 presents the calculation result of resistance and power versus speed of the monohull boat with Holtrop method of maxsurf software, where at speed 10 knots the resistance is 2.61kN and the power is 13.6kW, based on this is stated that the engine power of the monohull boat is 14kW.

Table 1: Main dimension of monohull boat

Displacement	3.051	tonne
Volume	2.977	m^3
Draft to Baseline	0.462	m
Immersed depth	0.462	m
Lwl	8.786	m
Beam wl	1.977	m
WSA	16.479	m^2
Max cross sect area	0.557	m^2
Waterplane area	15.148	m^2
Cp	0.608	
Cb	0.371	
Cm	0.61	
Cwp	0.872	
KB	0.328	m
KG	1.953	m

Table 2: Monohull boat resistance and power versus speed

Speed, kts	Resist, kN	Power, kW
0	0	0
1	0.01	0.01
2	0.05	0.06
3	0.09	0.14
4	0.18	0.39
5	0.38	1.03
6	0.55	1.7
7	1.11	4.08
8	1.69	7.19
9	2.07	9.56
10	2.61	13.6
11	2.97	17.2
12	3.17	19.6
13	3.47	23.4
14	3.79	27.8
15	4.03	31.1

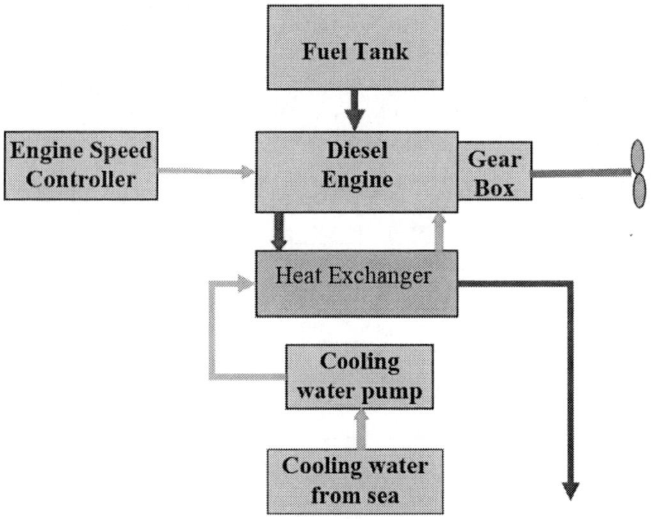

Figure 3: Diesel engine and propulsion system

2.2. CATAMARAN

At hybrid solar-diesel generator power system of a catamaran, where solar energy is transformed into electricity through photovoltaic cells (PV) runs the electric motor as prime mover on small catamaran passenger boats. Fig. 4 shows a small catamaran boat, where its body plan drawings is done with maxsurf software shown at Fig. 5. The displacement of the catatamaran is designed similar to the displacement of the monohull boat at Fig. 1.

Through software maxsurfpro is determined the main dimensions of the boat presented in table 3, then from the main dimensions is determined the deck area as passengers space where the length is 7.5m and breadth 3.5m, this area allows each passenger to occupy an area of about 1.05m2, which is twice than the area that is available for one passenger at monohull boat, so on catamaran boat the passengers can sit more conveniently.

Figure 4: Fore view and aft view of a small catamaran boat

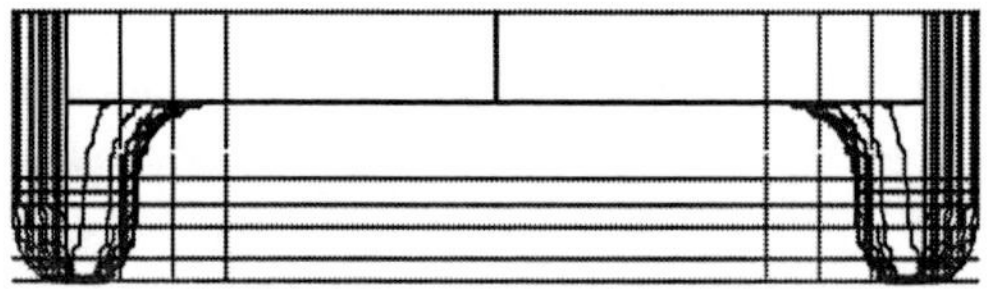

Figure 5: Body plan view of catamaran

2.2 (a) Water Resistance and Power

At Table 4 is presented the result of the water resistance and power versus speed of catamaran determined with maxsurf hullspeed software, where at speed 10 knots the boat resistance is 2.42 kN and power 12.62 kW ≈ 13 kW, which is about 7% less than of the monohull boat.

Table 3: Main dimension of catamaran boat

Displacement	3.047	tonnes
Volume	2.973	m^3
Draft to Baseline	0.51	m
Immersed depth	0.51	m
Lwl	9.273	m
Beam wl	3.85	m
WSA	23.164	m^2
Max cross sect area	0.423	m^2
Waterplane area	9.126	m^2
Cp	0.757	
Cb	0.491	
Cm	0.648	
Cwp	0.768	
KB	0.287	m
KG	2.15	m

Table 4: Resistance and power versus speed

Speed kts	Resist kN	Power kW
0	0	0
1	0.05	0.03
2	0.17	0.2
3	0.29	0.45
4	0.52	1.1
5	0.79	2.14
6	1.01	3.1
7	1.35	4.96
8	1.74	7.38
9	2.01	9.31
10	2.42	12.62
11	2.84	16.41
12	3.09	19.09
13	3.51	23.7
14	4.06	29.76
15	4.45	34.33

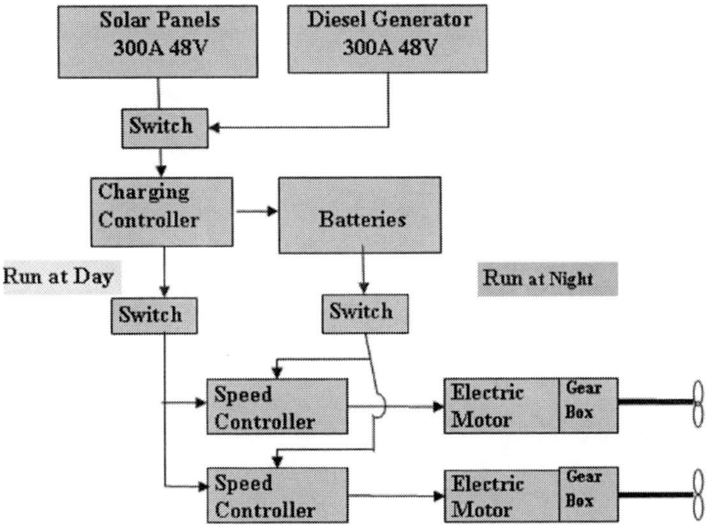

Figure 6: Hybrid solar-diesel generator electric propulsion system

2.2 (b) Hybrid Power System

Catamaran boats designed to connect closed islands generally are small in tonnage ferrying about 20 - 25 passengers similar passenger number as being carried by the monohull boats. The characteristics of these cat boats are suitable for the installation of hybrid solar-diesel generator energy system, due to enough area at the roof of the boat for installation of the solar cells panels. The propose power system components for these boats is shown in Fig. 6. The DC current produced at this system by the solar panels will run the electric motor before being channeled to the propeller. Part of the current is going to charge the batteries for running the boat during shortage of solar energy, in particular during the night when the capacity of batteries is low then the generator will perform its duty [3].

During the daytime the cat boat can use solar energy that could save fuel as by the monohull boats that is using diesel/gasoline engines, besides that the hybrid system has low operational cost and low maintenance/repair cost. Generally, solar panel has an operation life of 20 years, while E-motor has efficiency over 90 %. Under this arrangement, the propulsion system does not require thrust bearings as the electromotor will directly deliver the propeller thrust to the hull. Electric motors maintain high efficiency over a wide range of loads and speed. At 48 DC volts it will generate power up to 15 KW (≈ 20 HP). At low shaft speed and high torque they are capable using a simple, cheap and efficient transmission system in most applications. These motors are enable to get the best possible performance from the batteries or directly from the solar cells panels/diesel generator . High efficiency gives longer running time and less frequent battery replacement.

The generator and the solar cell panels have the same capacity that is 300 A, 48V DC, the battery capacity is 1000Ah.

3. RESULT AND DISCUSSION

This study focus is on implementing the alternative energy for powering small catamaran boat in place of monohull passenger boat where nowadays operating in these areas in a large number thus utilizing large quantities of fossil fuel. According to Muzathik et al. [4][5], the daily average sun shines hours are about 6-7 hours for the study area. In addition these vessels mainly operate during the day thus can benefit from solar energy for propulsion. Normally these vessels operate around 12 hours in a day, out of this about 6 hours power requirement should be derived using generator from 6.00 am – 9.30 am and from 3.30 pm - 6.00 pm where the solar radiation is low and is used only to charge the batteries and the other six hours it is ferrying by solar power. With this mode of operation the hybrid system could save fuel usage. This will also reduces the maintenance cost related to diesel-generator.

3.1 INCOME

Short distance routes between closed islands is assumed about 2 miles, which will be reached in 14 minutes by boat with speed 10 knots included the manuevring time. The boat will stay at the jetty for 16 minutes.Totally time for two ways trip and waiting time at the jetties will be 1 hour, the ferry boats will undergo about 12 trips daily. During 16 minutes waiting time at the jetty the number of passengers joined at each trip will be variously, assumed that the average number of passengers on one way trip is 50% of the boat capacity that is 10 people, so average total daily passengers during 12 trips is 12 x 2 x 10 passengers is 240 pass. If the boats will operate 300 days yearly and the the ticket price for one way trip is \$0.50, the income from ticket price for both types of boats will be 300 x 240 x \$0.50 = \$36,000 The income of hybrid cat boat is higher than of monohull boat income (see Fig. 8).

3.2 FUEL AND LUB OIL CONSUMPTION

Yearly fuel consumption of the mono-hull boat and the hybrid cat depends on the number of trip and number of yearly sailing hours of the boats. But for the hybrid cat boat the fuel consumption depends on the number of hours when the boat is powered by the diesel generator which are summarized in Table 5. The specific fuel consumption of small diesel engines and small generators are about 250g/kWh and specific lub oil consumption is 0.7g/kWh [6]. To obtain the daily fuel consumption it is depend on the engine power, specific fuel oil consumption (SFOC) and total daily sailing hours. The daily operational time of the mono hull and catamaran is 12 hours, but the cat boat will sail under solar power only about 6 hours daily that is from 09.30pm – 3.30pm and at the other 6 hours the cat boat will be powered by the diesel generator, from 6.00am – 09.30am and 3.30pm – 6.00pm, while during these hours the solar radiation is low and the electricity from the solar cells is used to charge the batteries [7]. If the cat boat is planning to operate at night, the diesel generator must always be on standby for charging the batteries and powering the boat, while at night the boat needs also lighting. About six hours daily the hybrid cat boat uses the solar power, that is why the sailing hours of the hybrid cat using diesel-generator is less than of the monohull, hence the yearly fuel and lub oil consumption of hybrid cat is less than of the monohull, the result is summarized in Table 5.

Table 5: Yearly sailing hours, fuel consumption and lub oil consumption for different trip number

Daily Trips	Yearly total sailing time, hours		Yearly fuel consumption, tonnes		Yearly lub oil consumption, liters	
	Monohull	Hybrid cat power by generator	Monohull	Hybrid cat power by generator	Monohull	Hybrid cat power by generator
4	560	0	2.45	0	7	0
5	700	0	3.07	0	9	0
6	840	0	3.68	0	10	0
7	980	140	4.29	0.48	12	1.18
8	1,120	280	4.91	0.96	14	2.35
9	1,260	420	5.52	1.44	15	3.53
10	1,400	560	6.13	1.93	17	4.70
11	1,540	700	6.75	2.41	19	5.88
12	1,680	840	7.36	2.89	21	7.06

3.3 MAINTENANCE COST

The average maintenance cost for diesel engine and diesel engine-driven generators is \$0.005-\$0.010 per kWh, according to the GTI (Gas Technology Institute), so this cost is depending on the power and the sailing hours of the boat during a year, hence the maintenance cost is increased by increasing of the trip numbers. In Fig. 7 is shown that the maintenance cost of the hybrid cat boat when ferrying daily (4 – 6) trips are zero because during these trips the diesel generator is not on duty, as the boat is powering by the solar cells panels while the solar radiation is high, but during trips more than six the generator must be in function due to not enough solar radiation (see Fig. 7).

3.4 PROFIT

Profit before tax is income – (operational cost + maintenance cost + crew salary). Hybrid cat boat has low operational and maintenance cost than of the monohull boat, hence its yearly profit is higher (see Fig. 8)

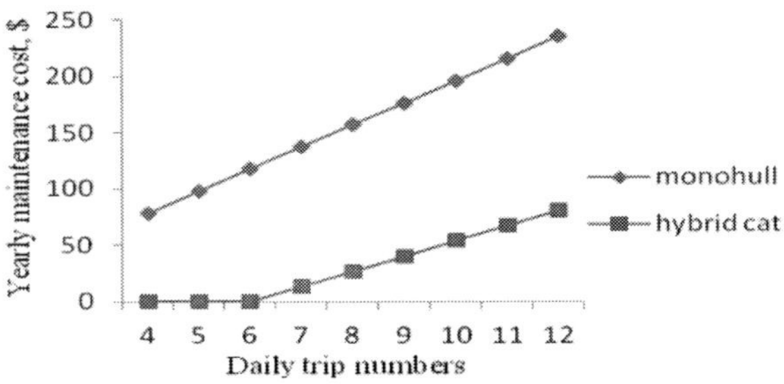

Figure 7: Yearly maintenance cost of monohull and catamaran

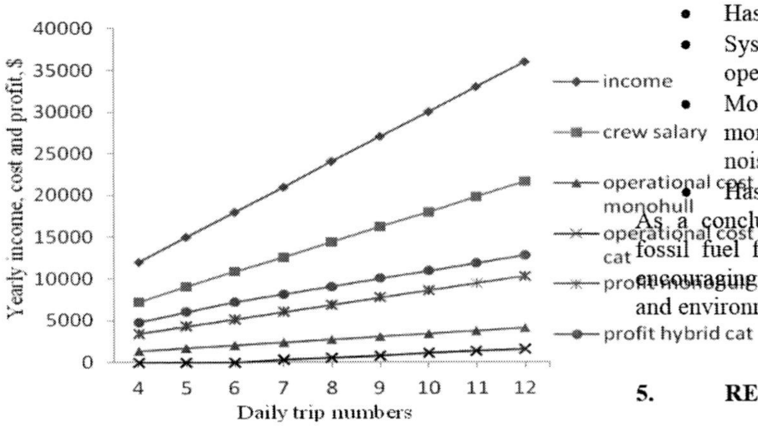

Figure 8: Yearly income, operational cost and profit of both boat types

3.5 CARBON EMMISION

With a good approximation we can say that burning a litre of diesel fuel oil produces 2.63 kg CO_2, the value depends on
the mixture of hydrocarbons and the engine, but does not vary much. The weight gains due to the fact that we have considered only the combustible, but we need to add oxygen in the combustion.
Formula for carbon emission is: CO_2 emission = Fuel Combusted (fuel units) * EF_{CO2} (kg CO_2/fuel units), where EF_{CO_2} is emission factor of diesel fuel oil the magnitude is 10.14 kg/gallon [8][9]
At Table 5 is presented yearly fuel consumption of hybrid cat boat which is more less than of the monohull boat, hence that lead to the larger occurrence of yearly carbon emission of the monohull boat compare to hybrid cat boat, which is shown in Fig. 9.

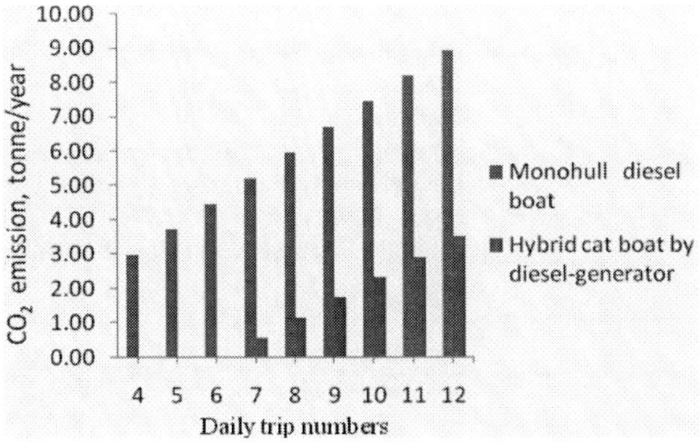

Figure 9: Yearly carbon carbon emission

4. CONCLUSIONS

Based on the study revealed that hybrid cat boat:

- Has less resistance and power.
- System need less fuel hence has less operational cost.
- More convenient for the passengers due to more space, good sight to the surroundings; no noise interference.
- Has less carbon emission, no waste water.

As a conclusion, using hybrid power system beside fossil fuel for small and medium catamaran boats is encouraging based on operation and operational costs and environmental issues.

5. REFERENCES

1. Zekai, S., Solar energy fundamentals and modeling techniques: atmosphere, environment, climate change and renewable energy, Springer, 2008

2. Comstock, J.P., Principle of Naval Architecture, SNAME, 1967.

3. Kevin Jeffery, Independent Energy Guide-electric Power for Home, Boat & RV, 2006.

4. Muzathik A.M., Wan Nik W.B., Samo K.B.and Ibrahim M.Z, Global solar radiation hourly estimate on horizontal plane, Journal of Physical Science, 21(2), pp. 51-64, 2010.

5. Kumar R. and Umanand, L., Estimation of global radiation using clearness index model for sizing photovoltaic system, Renewable Energy, 30 (15), pp. 2221-2233, 2005.

6. Marine Engine IMO Tier I Programme MAN Diesel, 2009.

7. O.Sulaiman, H.Aron, A.H.Saharuddin, W.B. Wan Nik, A.S.A. Kader, M.F. Ahmad (2011) Techno Economic Study of Potential using Solar Energy As a Supporting Power Supply for Diesel Engine for Landing Craft . International Journal of Business and Social Sciences, Vol. 2 No. 1

8. Dragan Ljevaja, Impact of Emissions of Marine Diesel Engines to Air Pollution on the Example of the Yugoslav River Shipping, World Transport Overseas Ltd., Milutina Milankovića 25b, 11000 Belgrade, Serbia, 2011

9. California Air Resources Board, Instructional Guidance for Mandatory GHG Emissions Reporting, December 2008

6. AUTHORS' BIOGRAPHY

Ferry Manuhutu, contract lecturer at University Malaysia Terengganu(UMT), Malaysia. Lecturing: Naval Architecture and Ship Construction, Shiprepair and maintenance system. Supervising final year projects of undergraduate students (B.Appl Sc): Feasibility study of wind and solar hybrid power system of catamaran for east coast Malaysia.

W.B. Wan Nik, Associate professor in Marine Corrosion, dean of faculty of Maritime Studies and Marine Science UMT, Malaysia. Lecturing: Power Plant and Ship Machinery, Marine Technology and Environment. Journal publication: W.B. Wan Nik, M.Z. Ibrahim, K.B. Samo, A.M. Muzathik (2012), Monthly mean Hourly Global Solar Radiation Estimation, Solar Energy, V86, I1, January 2012, p379-387

M.F. Ahmad, Associate professor in Coastal Engineering, head of department of Maritime Technology UMT, Malaysia. Lecturing about Marine technology system, Research Method in Technology. Research completed: Techno Economic Study of Potential using Solar Energy As a Supporting Power Supply for Diesel Engine for Landing Craft . International Journal of Business and Social Sciences, Vol. 2 No. 1

D. Pelupessy, lecturer at Department of Mechanical Engineering PattimuraUniversity. Lecturing: Marine Automation and Control, Maritime Engineering Design. Martec 2008 proceeding: Catamaran powered by solar energy for interisland transportation in the Maluku province.

WME. Wattimena, lecturer at Department of Mechanical Engineering Pattimura University. Headof laboratory of maintenance and repair of engines. Lecturing subjects: Ship maintenance and repair, Ship engine and machinery system. Proceeding sail Banda: Hybrid solar-wind-diesel power plant for small islands in Maluku province.

THE POWERING OF FUTURE SHIPS TAKING INTO CONSIDERATION ECONOMIC VIABILITY AND ENVIRONMENTAL ISSUES

I.K.A.P. Utama, ITS, Surabaya, Indonesia and **A.F. Molland,** University of Southampton, UK

SUMMARY

Economic factors such as changing fuel costs and environmental issues such as emission of greenhouse gases, pollution, wash and noise, all ultimately lead to the need to minimise ship propulsive power. The problem of minimising power can be addressed at the design stage and during operation.

The various requirements and needs that impact upon the current and future design and operation of ships are described. The main components of the powering process are reviewed. Individual components and power saving methods are identified which should enable improvements in the overall efficiency of propulsion to be made.

Potential levels of savings in propulsive power at the design stage and during operation are described and discussed. Appropriate design methodologies and procedures, taking into account economic and environmental factors, are proposed for the design of future ships.

NOMENCLATURE

A	Propeller blade area (m^2)
a	Axial inflow factor
a'	Circumferential (rotational) inflow factor
B	Breadth (m)
BAR	Blade area ratio
C_B	Block coefficient
C_T	Propeller thrust coefficient [$T/\frac{1}{2}\rho A V^2$]
D	Propeller diameter (m)
EEDI	Energy efficiency design index
J	Propeller advance ratio [Va/nD]
K_T	Propeller thrust coefficient [$T/\rho n^2 D^4$]
L	Length (m)
LCB	Longitudinal centre of buoyancy
n	Propeller rate of revolution (rps)
NPV	Net present value
P	Propeller pitch (m)
P_D	Delivered power (kW)
P_E	Effective power (kW)
R	Resistance (kN)
RFR	Required freight rate (cost/tonne)
sfc	Specific fuel consumption (kg/kW.hr)
T	Thrust (kN)
t	Thrust deduction factor
V	Ship speed (m/s)
Va	Propeller advance speed (m/s)
w_T	Wake fraction
η_D	Quasi propulsive coefficient
η_H	Hull efficiency
η_0	Open water efficiency
η_R	Relative rotative efficiency
η_a	Ideal efficiency
η_r	Rotational efficiency
η_f	Frictional efficiency
∇	Displacement volume (m^3)
ρ	Density, sea water (kg/m^3)

1. INTRODUCTION

1.1 BACKGROUND

The design of merchant ships has traditionally been centred on economic viability. The main economic drivers amount to the construction costs, disposal costs, ship speed and, in particular, fuel costs. These need to be combined in such a way that the shipowner makes an adequate rate of return on the investment. More recently, attention has been drawn to the need to take environmental issues into account. The main environmental drivers amount to emissions, in particular the need to reduce the emission of CO_2, pollution, noise, anti-foulings and wave wash. Economic and environmental pressures thus combine to create the need for a fresh appraisal of the estimation of ship propulsive power and the choice of suitable machinery. Both issues present the need to minimise propulsive power. Minimising propulsive power can be addressed at the design stage in terms of hull form design and propulsor design, and during operation which can include the use of suitable changes in trim, route changes and slow steaming. The reduction in wave wash and the developments in anti-foulings are described in references such as [1], [2] and [3]. This paper concentrates on the reduction of CO_2 emissions.

1.2 QUANTIFYING ENVIRONMENTAL IMPACT

The emissions from ships include NOx, SOx and CO_2, a greenhouse gas. CO_2 emissions have a global climate impact and a concentrated effort is being made worldwide towards their reduction. In order to monitor and quantify CO_2 emissions, the International Maritime Organisation (IMO) has developed an Energy Efficiency Design Index, IMO [4], [5].

The general form of the index, as proposed by IMO, is as follows:

$$\text{EEDI} = \frac{P \times sfc \times C_F}{C \times V} \text{ gm } CO_2\text{/tonne.mile} \qquad (1)$$

where P is power (kW), sfc is specific fuel consumption (gm/kW.hr), C_F is a CO_2 conversion (tonne CO_2/tonne fuel), C is the capacity of the ship (deadweight tonnes or Gross Tonnage) and V the speed (nautical miles /hr (knots), or km/hr). As such, the Energy Efficiency Design Index can be seen as a measure of a ship's CO_2 efficiency. This is very much the general, or generic, form of the equation as the power will be made up of the propulsive and auxiliary power, the capacity C of the ship will in the main be deadweight tonnes; passenger ships will use gross tonnage. Speed has to be clearly defined as it could be taken as the design speed, or some average speed expected in operation. Similarly, power may be the design calm water propulsive power, or power taking into account average increases due to weather.

When considering the overall form of the Energy Efficiency Design Index it is clear that in order to reduce the index for a given ship at a given speed, a decrease in propulsive power must be achieved and/or improvements made in engine efficiency with a reduction in sfc.

For explanatory and comparative purposes, this paper uses the general form of the index, Equation (1), with P as the service propulsive power, capacity C as deadweight tonnes and V the service speed in knots. The example cargo ship shown in Table 1 with assumed sfc = 190 gm/kW.hr and C_F = 3.17 tonne CO_2/tonne fuel (IMO [4]) would have a Energy Efficiency Design Index = 12.89 gm/tonne.mile.

Table1: Particulars of example cargo/container ship

L	145 m
B	24 m
T	10.7 m
C_B	0.710
Δ	27100 tonnes
Deadweight	20000 tonnes
V	16 knots
Ps service power	6850 kW

As there are proposals to introduce a form of CO_2 emissions control, a limit will be set on the Energy Efficiency Design Index for new builds which will come into force in 2013. This immediately sets great importance on the specific definition of each of the components in Equation (1), Reference [4], together with future design procedures and decisions.

1.3 OBJECTIVES

The objectives of the paper are, first, to review the main components of power and identify where improvements can be made and, second, to develop an awareness and design methodology that will take account of economic and environmental issues.

2. POWERING

2.1 OVERALL CONCEPT

The overall concept of the powering system may be seen as converting the energy of the fuel into useful thrust (T) to match the ship resistance (R) at the required speed (V), Figure 1.

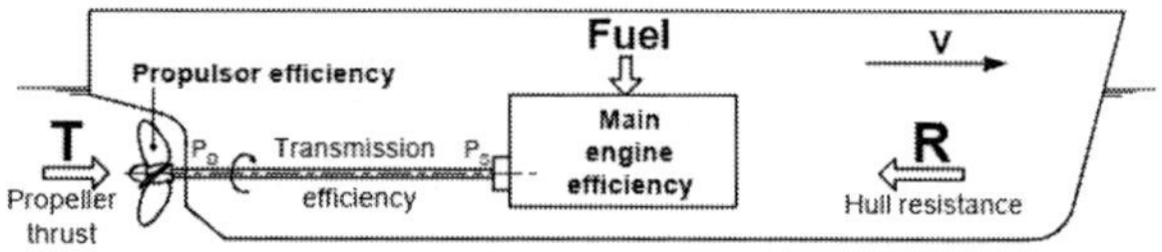

Figure 1: Overall concept of energy conversion

It is seen that the overall efficiency of the propulsion system will depend on:
- fuel type, properties and quality;
- the efficiency of the engine in converting the fuel energy into useful transmittable power; and
- the efficiency of the propulsor in converting the power (usually rotational) into useful thrust (T).

The present study concentrates on the performance of the hull and propulsor, primarily considering, for a given set of constraints, how resistance (R) may be reduced and thrust (T) may be increased. Available engines are summarised later. Detailed accounts of propulsion engines may be found in sources such as Woodyard [6] and Molland [7].

2.2 COMPONENTS OF POWERING

The main components of powering are identified and summarised. This allows assessments to be made of the areas where changes and potential improvements may be made.

2.2(a) Propulsive power

The power delivered to the propeller, delivered power (P_D), may be defined as:

Delivered power (P_D)

$$= \frac{\textit{Effective power } (P_E)}{\textit{Quasi propulsive coefficient } (\eta_D)} \qquad (2)$$

$$\text{Effective power } (P_E) = R \,.\, V \qquad (3)$$

where R is total resistance (kN) of the naked hull and appendages, together with above water air drag of the hull and superstructure. V is ship speed (m/s).

The total naked hull resistance is made up of friction, viscous pressure (or form) and wave components, as shown in Figure 2. These basic hull components are applicable to displacement ships and most semi-displacement ships. For faster vessels, other components arise such as transom, spray and induced drag. The ships discussed in the current work are mainly single screw and appendage drag is generally relatively small. Air drag may be significant and is discussed later.

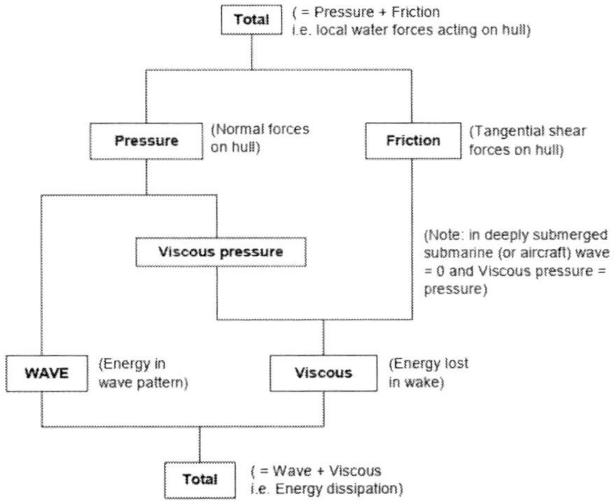

Figure 2: Basic resistance components

2.2(b) Propeller efficiency

The components of quasi propulsive coefficient (η_D) may be written as:

$$\eta_D = \eta_0 \cdot \eta_H \cdot \eta_R \tag{4}$$

where η_0 is the open water efficiency of the propeller, η_H is the hull efficiency and η_R is the relative rotative efficiency.

η_R takes account of the differences between the propeller in the open water condition and when behind the hull, and lies typically between 0.98 to 1.02.

η_H takes account of the interaction between the hull and propeller and is defined as:

$$\eta_H = \frac{(1-t)}{(1-w_T)} \tag{5}$$

where t is the thrust deduction factor and w_T the wake fraction. η_H lies typically between 1.0 and 1.25 for displacement ships. The formula indicates how changes in thrust deduction (t) due, for example, to the presence of a rudder or other device will influence overall propeller efficiency. Similarly, the influence of wake fraction (w_T) can be seen and quantified.

η_0 is the open water efficiency of the propeller and will depend on the propeller parameters, as discussed later in Section 3.

3. POTENTIAL SAVINGS IN POWER

3.1 DESIGN STAGE

3.1.1 Hull form

Hull resistance is dominated by the principal hull parameters such as $L/\nabla^{1/3}$, C_B, B/T and LCB. Local detail, such as the use of 'V' or 'U' shaped sections forward and/or aft, will have an effect, as will the use of bulbous bows. The use of bulbous bows should be made with caution in that there are relatively specific areas where they can be used to advantage. Suitable areas for the application of bulbous bows can be found in [8] and from the work of Kracht [9]. It is found that the choice, and cost, of employing a bulbous bow is design specific. Hochkirch and Bertram [10] report on an investigation into the design and re-design of bulbous bows for container ships when slow steaming. Vortex generators can be employed to re-align the aft end flow and delay separation. This is often done to provide a cleaner flow into the propeller, rather than necessarily reducing resistance, Anon [11].

3.1.2 Hull surface finish

Hull surface finish is fundamental to the levels of hull skin friction resistance. Much research has been carried out to demonstrate the benefits of a good surface finish, for example Townsin et al. [12], [13], [14]. The frequency of docking to clean the hull has normally been assessed on economic grounds. If an emissions trading scheme, or some form of carbon 'subsidy' were introduced, then the emphasis might change to a reduction in power, driven by a reduction in CO_2 emissions, rather than solely for economic purposes.

3.1.3 Appendages

Appendages, such as bilge keels, shaft brackets and rudders require careful design. This might entail flow visualisation tests or CFD investigations to optimise the alignment of bilge keels and shaft brackets. Rudders should be considered as part of the propeller-rudder combination in respect of thrust deduction and propulsive efficiency changes, Molland and Turnock [15], Anon [11].

3.1.4 Air Drag

Air drag of the above water hull and superstructure is generally a relatively small proportion of the total resistance for tankers, bulk carriers and container ships. However, for a large vessel, any reductions in air drag may be worth pursuing. Air drag values for commercial ships can typically be found in Isherwood [16], van Berlekom [17], Gould [18] and Molland and Barbeau [19].

3.1.5 Propulsive efficiency

3.1.5(a) Individual components of propeller efficiency:

For a fixed set of propeller parameters, η_0 can be considered as being made up of:

$$\eta_0 = \eta_a \cdot \eta_r \cdot \eta_f \quad (6)$$

where η_a is the ideal efficiency, η_r accounts for losses due to fluid rotation induced by the propeller and η_f accounts for losses due to blade friction drag, Dyne [20], [21], Molland et al [22].

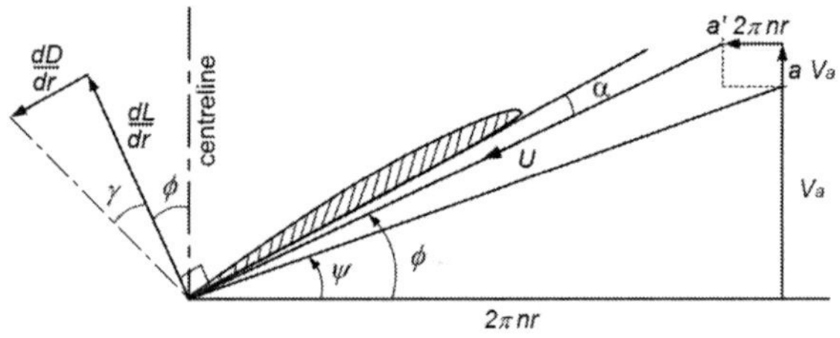

Figure 3: Blade element diagram

An investigation has been carried out to determine the likely values of these three components of efficiency. Blade element momentum theory was used, [22], [23] and, based on the blade element diagram, Figure 3, it can be shown that:

$$\eta_a = \frac{1}{(1+a)} \quad (7)$$

$$\eta_r = (1-a') \quad (8)$$

$$\eta_f = \frac{\tan\phi}{\tan(\phi+\gamma)} \quad (9)$$

where a and a′ are the axial and rotational inflow factors, derived from momentum considerations and corrected for finite number of blades using Goldstein correction factors [22].

The investigation used a propeller with a pitch ratio P/D = 1.0, BAR = 0.700 and 4 blades for a range of J values, hence thrust loading, C_T. The thrust loading coefficient C_T is defined, and related to K_T/J^2, as follows:

$$C_T = \frac{T}{0.5\rho\frac{\pi D^2}{4}Va^2} = \frac{K_T}{J^2}\cdot\frac{8}{\pi} \quad (10)$$

For a fixed pitch, decrease in J leads to an increase in thrust loading K_T/J^2, or C_T.

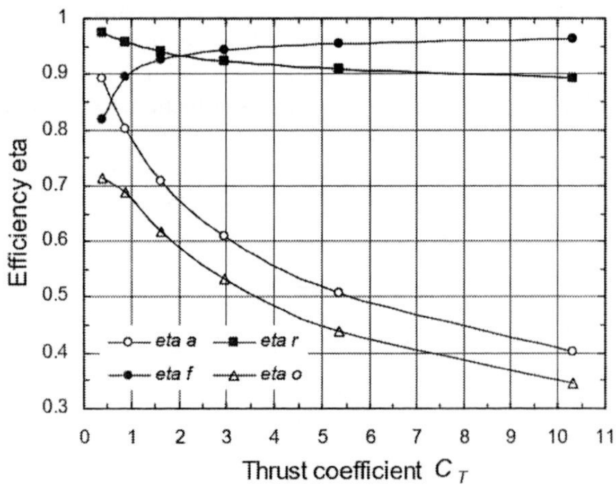

Figure 4: Components of propeller efficiency

The results of the investigation are shown in Figure 4. It is seen that at say C_T = 0.86 (J = 0.75), η_a = 0.80, η_r = 0.96 and η_f = 0.90. With increase in thrust loading up to C_T = 10.3 (J = 0.35), it is seen that η_f increases from about 0.82 to 0.96, η_r decreases from about 0.97 to 0.89 and η_a decreases significantly from 0.89 to 0.40. Overall, η_0 decreases from about 0.70 to 0.35.

This breakdown of the individual components of η_0, Equation (6), is important as it indicates where potential savings might be made. The following comments are made on practical methods used to improve the efficiency of each component:

(i) η_f Savings in friction can be achieved by using the minimum blade thickness consistent with strength considerations and minimum blade area consistent with cavitation requirements. Reduction in friction can also be achieved by decreasing the local inflow velocity (by reducing revolutions) with an appropriate increase in diameter. Friction loss may also be reduced by reducing revolutions and increasing pitch, although this will lead to increased rotational losses which may have to be compensated by stators downstream. Surface finish influences friction, particularly during operation.

(ii) η_r Recovery of rotational losses can be achieved in various ways. The most effective way is to employ contra-rotating propellers, where much of the flow rotation loss can be removed. The most common way is by the use of pre and post swirl stators. The rudder (downstream of a propeller) acts as a post swirl stator, but also blocks the flow, resulting in pressure (axial) losses which can negate much of the rotational savings. Pre-swirl (upstream) stators can entail fins, and post swirl downstream using fins attached to the rudder. Asymmetric sterns can be employed which put pre-rotation into the propeller inflow. It should be noted that pre and post swirl fins and asymmetric sterns can increase the resistance, hence reducing some of the effective savings in propeller efficiency.

(iii) η_a Axial losses are by far the largest, Figure 4. Theory and practice indicate that an increase in diameter with commensurate changes in P/D and rpm will lead to improvements in axial efficiency (together with some improvement in frictional efficiency). Maximising diameter is therefore of fundamental importance. Propeller tip clearances will normally limit the maximum diameter. Accelerating ducts may be used to advantage in conditions of high thrust loading, although duct friction will tend to remove any savings at moderate or low thrust loadings. An upstream semi-duct has been employed to improve the axial efficiency by directing part of the frictional wake to inside the propeller diameter.

3.1.5(b) Effect of propeller diameter

In order to quantify the influence of propeller diameter on propulsive efficiency, a practical study was carried out on a small cargo/container ship with a length of 155m, a design load draught of 10.5m and a service speed of 18 knots. A survey of container ships indicated that the choice of propeller diameter was between 65% and 74% of the design load draught. The lower values are presumably applied where operation is expected at draughts significantly less than the design load case. This can create a significant power penalty. For example, when this range of diameters is applied to the 155m vessel, then the propeller diameter would be between 6.8m and 7.8m. If this were transformed into propeller efficiency improvements, then the order of increases are shown in Figure 5, with the propeller efficiency η_D, Equation (4), improving from 0.646 to 0.678, an improvement of some 5%. To be able to incorporate such a large increase in diameter is unlikely in a particular design situation, but the attraction of applying any increase in diameter is apparent.

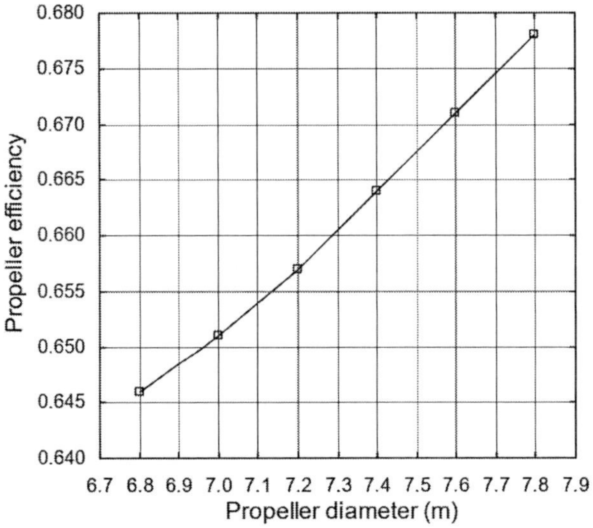

Figure 5: Change in propeller efficiency η_D with change in propeller diameter

3.1.5(c) Alternative propulsors

Alternatives to the simple solid fixed-pitch propeller may be summarised as ducted, controllable pitch, contra-rotating, cycloidal and podded units. Generally, these are employed for specific applications where improvements in propulsive efficiency can be made, noting that for ocean-going merchant ships such units must be robust, reliable and safe. There are also a number of detailed modifications that can be made to the simple propeller, including the use of tip fins and tip rake. Based on the need for robustness and reliability, the alternative propulsors discussed are unlikely to have a significant impact on overall emissions reduction for large ships such as tankers, bulk carriers and container ships.

3.1.5(d) Inclined keel

The need to reduce power and emissions would suggest that it should be worthwhile to investigate further the inclined keel concept. In this case, the keel is inclined (equivalent to designing in trim) and a significantly larger propeller can be employed, [24], [25]. This is similar to the approach used for tugs and trawlers. In the case of a larger vessel, such as a container ship, the draught amidships would be the design draught and the ship would ballast back to level keel if required by port draught limitations. As an example, if the 145m cargo/container ship in Table 1 had a 2.0m trim by the stern, and assuming this could be transformed (by the redesign of the aft end) into an increase of 1.0m in propeller diameter then, using Figure 5, this would suggest an expected increase in propeller efficiency of about 5%. There may be some increase in resistance with an inclined keel, but the indications are that the gains to be made from the increased propeller diameter are greater than the losses due to the increase in resistance. These findings would suggest that the concept deserves further consideration.

3.1.6 Propeller-hull interaction

3.1.7

This in an area that can have a significant effect on the overall propulsive efficiency. For example, examination of Equation (5) indicates how thrust deduction (t) and wake fraction (w_T) affect hull efficiency, whilst Equation (6) includes the rotational losses η_r for the propeller which might be recovered by the application of suitable devices. The propeller-hull interaction is dominated by propeller hull clearances and aft end hull shape, for example fineness of waterline endings, depending on C_B and *LCB*, and/or the use of 'U' or 'V' sections upstream of the propeller. This is modified by the possible presence of shaft brackets and rudders. Fundamental also is propeller rudder-interaction, having an influence on thrust deduction and some recovery of propeller induced rotation of the flow. A further basic change would be the use of a 'bulbous' stern, with or without asymmetry. Beyond these fundamental aspects are detailed devices that can contribute to improvements in efficiency. These include vortex generators, which are claimed to have led to 4-6% reduction in fuel consumption, Anon [11], and a duct upstream of the propeller, which is claimed to save up to 4% of power for large full form vessels, Anon [26]. Savings of between 2-4% might be expected from the application of pre and/or post swirl stators. An integrated twisted rudder, bulb and propeller hubcap is described in Anon [27] and it is suggested that savings in power of up to 10% might be achieved with careful integrated design of hull, propeller and rudder.

3.1.8 Propulsion machinery

The main propulsion machinery is responsible for converting the energy in the fuel into useful mechanical power, Figure 1. The main types of engine, suitable for the propulsion of commercial ships may be summarised as:

- Low, medium and high speed diesel engines;
- Gas turbines;
- Electric motor, inboard or within a podded drive.
- Steam turbines.

The principal properties of the various propulsion engines, such as size, mass, fuel consumption and emissions are described in some detail in Woodyard [6] and Molland [7]. It should be noted that engine manufactures have made significant improvements in overall engine efficiency in recent years, leading to reductions in the fuel consumption and emissions.

3.1.8 Alternative fuels

A number of alternative fuels are under consideration which would reduce the emission of greenhouse gases and reduce the dependence on oil, ECSA [28]. These include bio fuels, nuclear power, LNG and fuel cells. Bio fuel does not contain sulphur and reduces the emission of CO_2. It does, however, have a high price and may not be available in suitable quantities for shipping. Nuclear power has a proven track record for naval ships and icebreakers. The mass and size of nuclear units has decreased significantly and the application of nuclear power to merchant ships is currently being studied, [29]. The use of LNG would reduce CO_2 emissions. The large volume of stowage required for LNG tends to make it less viable for large ocean-going ships, although it has several suitable applications for small ships. Fuel cells may become viable in the future, but at present energy efficiency levels are not suitable for the propulsion of large ships.

3.2 POWER SAVINGS DURING OPERATION

3.2.2 Speed

For most displacement ships, power varies approximately as speed cubed. Any reduction in speed can therefore offer significant reductions in power and the emission of greenhouse gases. On an economic basis, the reduction in speed leads to a saving in fuel but a loss of earnings and there is a fine balance between them to derive the 'optimum economic' speed. If the physical changes are examined, it is found that, initially, starting from a low speed, as speed is increased the increase in earnings increases at a greater rate than the power and fuel. This continues until a speed is reached when the increase in fuel is greater than the increase in earnings. It is found that optimum speed decreases with increase in fuel costs and provides the reason for the use of lower speeds, or slow steaming, in periods of high fuel costs. It is important to note that, as power varies (approximately) as speed cubed, a reduction in speed will lead to a reduction in the in the EEDI (CO_2 index), Equation (1). For example, if the speed of the ship in Table 1 is reduced from 16 to 14 knots, then there is a significant reduction in the EEDI from 12.89 to 9.87 gm/tonne.mile. Decisions on levels of speed reduction are also likely to depend on the overall operation of a number of ships to transport a certain amount of cargo.

3.2.2 Effects of trim on hull resistance

Merchant ships are normally designed for level trim in the load condition and some trim by the stern in ballast condition. This will normally ensure adequate propeller immersion in the ballast case together with forefoot immersion The effects of trim on hull resistance have been investigated using data from the BSRA Series [30]. The results are for the standard BSRA vessel with L = 122m and B = 16.8m. The ballast condition was tested at a draught equivalent to 4.9m both at level keel and with a trim of 2.4m by the stern. The results over a range of speeds for three block coefficients are shown in Figure 6. It is seen from Figure 6 that significant changes in hull resistance can occur with change in trim. In the case of the C_B = 0.65 vessel, the major changes occur at higher speeds, where up to 3.5% decrease in resistance is observed. The C_B = 0.70 case shows up to 5% reduction in resistance at higher speeds. For C_B = 0.75, the resistance in the trimmed condition is up to 6% higher at low speeds and up to 3% lower at higher speeds. The results in Figure 6 serve to illustrate levels of change in hull resistance with change in trim. At the same time, change in trim can change hull efficiency factors such as wake fraction, with consequent change in overall propulsive efficiency. An overall investigation should take such effects into account. The effects of trim on hull resistance for a particular vessel can be determined at the model tank testing stage, and/or by the use of CFD, which can provide guidance to the ship operator. Alternatively, investigations can be carried out on the ship in service by monitoring say power or fuel consumption (F) over a period of time for different amounts of trim. In a manner similar to voyage analysis [22], changes due to trim might be monitored using a factor F_C, such as described in Equation (11), which can also correct for any changes in displacement (Δ) and speed (V).

$$F_C = \frac{\Delta^{2/3} V^3}{F} \tag{11}$$

Hansen and Freund [31] describe a detailed study into the derivation of the effects of trim and application to ships in operation.

3.2.3 Weather routeing

This is now a well practised procedure by many shipping companies. It entails trading a relative decrease in fuel consumption for an increase in distance to travel around bad weather. To work effectively, a knowledge is required of the performance of the ship in a seaway, in particular, speed losses in the various forecasted sea conditions. Such procedures are, for example, described by Satchwell [32]. The practice should lead overall to the emission of less greenhouse gases.

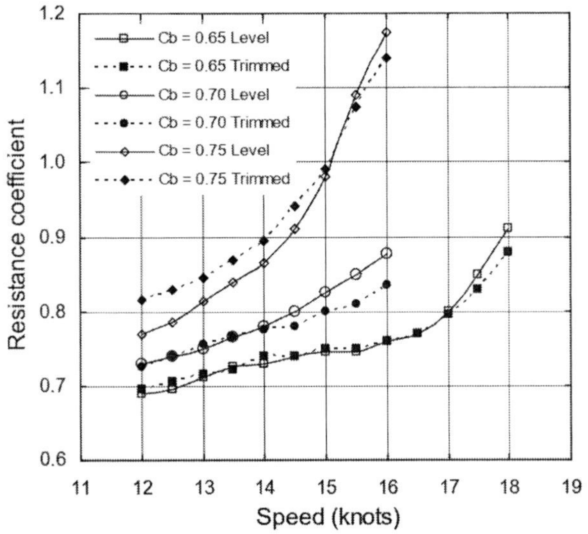

Figure 6: Effect of trim on resistance

3.2.4 Hull coatings

A smooth surface, with low roughness, will normally lead to lower frictional and viscous pressure resistance. From a hydrodynamic point of view, the underlying objective is to provide a smooth hull surface finish when the vessel is constructed and to maintain a clean smooth surface in service. Research by Candries and Atlar [3] indicates that reductions in skin friction resistance may be achieved with foul release coatings.

3.2.5 Hull/propeller cleaning

Hull cleaning is known to decrease overall power, but has usually been carried out on a strictly economic basis, see Townsin et al. [12], [13], [14]. The decrease in CO_2 emissions, and possible emissions trading for increases in maintenance costs, could provide the operator with the incentive to clean the hull and propeller over shorter intervals of time.

3.3 AUXILIARY PROPULSION DEVICES

There are a number of devices that provide propulsive power using renewable energy. The energy sources are wind, wave and solar. Devices using these sources are described in references such as [33], [34], [35], [36], [37] and [38].

Whilst a number of the devices may be impractical as far as propulsion is concerned, some, such as wind turbines and solar panels, may be used to provide supplementary power to the auxiliary generators. This will lead to a decrease in *overall* power (propulsion and auxiliary electrical generation) and an overall reduction in emitted greenhouse gases.

4. ECONOMIC AND ENVIRONMENTAL ISSUES

The factors driving current research and investigation into improving the overall efficiency of propulsion of ships are both economic and environmental. Fundamentally, improvements in efficiency of propulsion should lead directly to improvements in the economic return and a decrease in greenhouse gas emissions. This means there is a double incentive to pursue such efficiency improvements. There are, however, some possible technical changes that will decrease emissions, but which may not be economically viable. For example, the use of controllable pitch propellers (for off-design improvements) or contra-rotating propellers to improve efficiency and decrease emissions, will both incur increases in first and maintenance costs, detracting from the economic efficiency. Many of the auxiliary powering devices, using renewable energy sources and enhanced hull coatings, are likely to come into this category. There are suggestions that emissions trading for ships or, in effect, some form of 'subsidy' for decreasing emissions, may be introduced in the future. If this is the case, all means of improvement in powering and reduction in greenhouse gas emissions should be explored and assessed, even if such improvements may not be directly economically viable. Studies have been carried out over many years to derive the most suitable combination of hull parameters for a particular vessel at a particular speed with a given fuel cost. Analysis is usually based on some economic measure of merit, such as NPV, yield or required freight rate. The resulting dimensions will depend on speed, build cost and fuel cost. For example, if speed is reduced, the vessel will tend to be shorter and construction costs will reduce, whilst for higher fuel costs, optimum length and L/B tend to be larger, the decrease in power and fuel offsetting the increase in build cost. Parametric changes in main hull dimensions for tankers have been carried out by a leading oil company, providing indications of what savings in power might be achieved, SSPA [39].

A study has been made into the effects of parametric changes in hull parameters for the small cargo/container ship described in Table 1. This entailed running the ship at 16 knots with 18 voyages/year and changing *L/B* ratio in a methodical manner. The ship design software ShipDes, described by Molland et al. [38], was used for the investigation. The results are shown in Figure 7. Changes in the construction costs/annual charges and fuel cost changes follow expected trends, namely, as L/B increases, construction costs and annual charges increase and fuel costs decrease. Observation of the RFR results in Figure 7 indicate that with fuel at $150/tonne the *L/B* should be about 6.5. Figure 7 also illustrates how higher

L/B leads to a lower CO_2 index (EEDI). However, above L/B of about 6.5, the RFR tends to increase, leading to a decrease in economic efficiency. This illustrates the conflicting demands of environmental efficiency and economic efficiency. What is important to note is that, with high fuel prices, and pressure to reduce emissions and power, it may well be necessary to move to higher *L/B* ratios than is currently the practice. On the same basis, suitable values for the other hull parameters, such as *B/T*, C_B and LCB, should also be re-visited. It is apparent that the combined influences of fuel costs *and* CO_2 emissions are likely to take a more important role in the choice of the overall hull parameters for future tonnage.

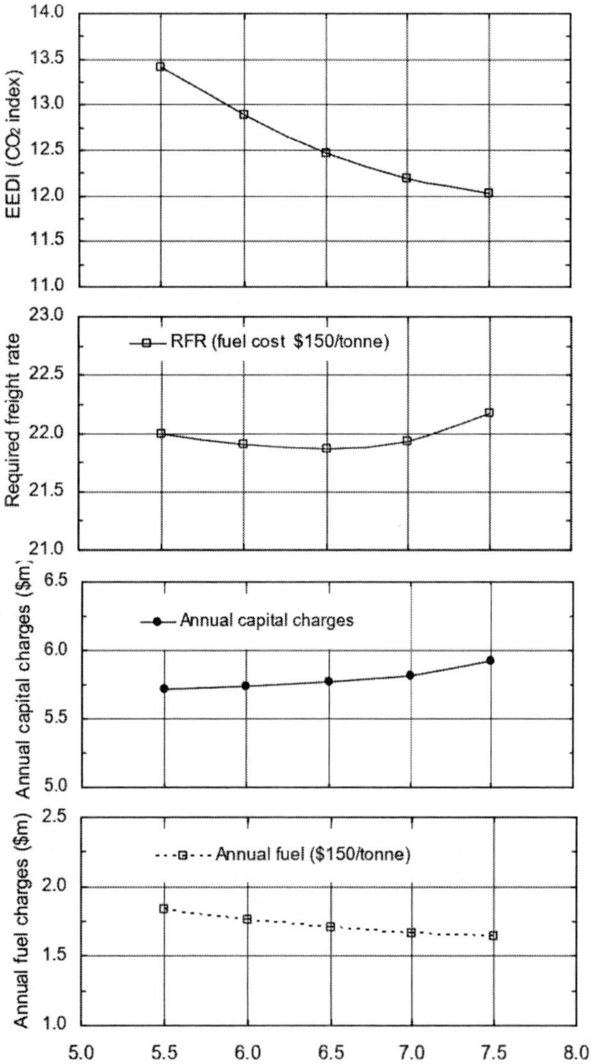

Figure 7: Influence of L/B on Annual charges and fuel, Required Freight Rate and CO_2 index

5. DESIGN PROCEDURES

It is clear from the earlier Sections that procedures for quantifying the impact of greenhouse gas emissions need to be incorporated in the ship design process. This could, for example, entail the incorporation of the EEDI (Equation (1)) as an objective function. Figure 8 shows a traditional ship design approach where the objective function is some economic criterion such as NPV or RFR, Molland [7], Schneekluth and Bertram [40].

Figure 9 indicates how the environmental effects may be incorporated in the ship design process. The use of such an approach allows design changes, technical innovation and auxiliary power devices to be incorporated in the feasible technical design, and a cost benefit analysis of these changes carried out in the usual way, Schneekluth and Bertram [40]. Thus the objective function for optimising on an economic basis

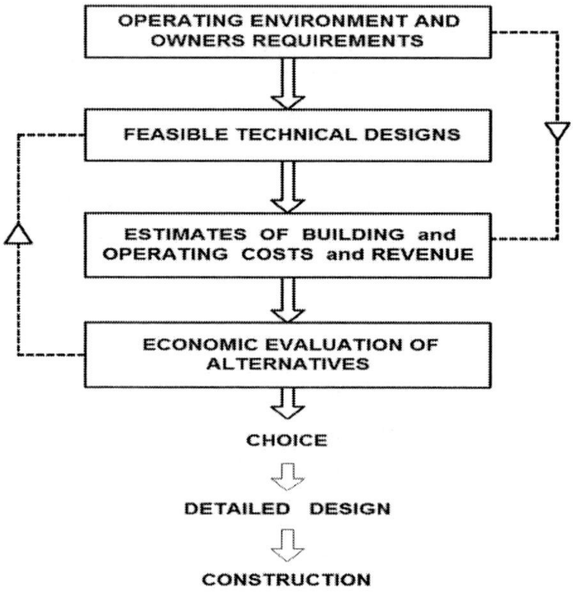

Figure 8: Overall flow path

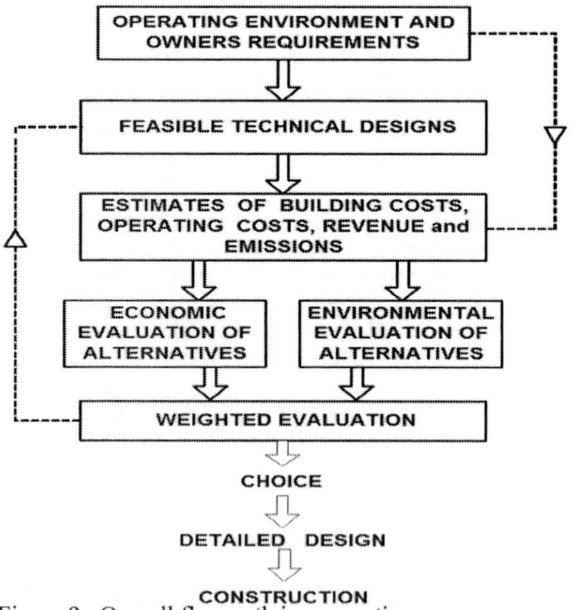

Figure 9: Overall flow path incorporating environmental effects

might be NPV or RFR, whilst the environmental 'optimum' might be to achieve the lowest EEDI. Earlier examples, such as that illustrated in Figure 7, have indicated that the economic and environmental optima may not coincide.

The design path now becomes a multiple criteria problem, see for example Sen [41] and Schneekluth and Bertram [40]. Weightings will have to be applied depending on what financial incentives might be given, directly or indirectly, to arrive at an environmental optimum which is not necessarily the economic optimum. The weightings are likely to depend on fuel cost levels and incentives in carbon trading schemes.

6. CONCLUSIONS

6.1 General

A number of areas have been identified where initial design changes and investment at the construction stage can lead to significant savings in fuel consumption and emission of greenhouse gases. Changes may be made at the design/construction stage, or modifications carried out whilst the ship is in service. Whatever changes are made, these must be robust, reliable and safe.

6.2 Resistance

Several areas have been identified where overall propulsive efficiency may be improved. In terms of resistance, optimisation of overall hull shape parameters should be investigated, together with attention to the fore end in terms of bulbous bows and local section shape, and to the aft end in terms of section shape and the interaction of the wake with the propeller. CFD may be usefully employed to develop suitable hull shapes for particular operational conditions.

6.3 Propeller efficiency

Propulsive efficiency offers a number of areas for improvement. Maximising and optimising propeller diameter is fundamental and there are some opportunities for doing this. Attention to local detail can be productive, including the hull to propeller and propeller to rudder interactions where efficiency gains can be made. These include asymmetric and bulbous sterns. CFD should be used to further complement model tests for hull-propeller-rudder interaction effects, where worthwhile improvements in overall efficiency of the propulsion unit might be achievable. At a greater level of detail, boss cap fins, upstream ducts, pre and post swirl devices and twisted rudders can be used to advantage.

6.4 Savings

In reacting to the pressure to reduce propulsive power, the designer will need to investigate every feasible possibility. This might entail deriving small improvements from a number of the component parts which, collectively, will provide worthwhile savings in overall power and a reduction in the emission of greenhouse gases.

6.5 Economic viability

With the increased pressure from the environmental point of view and with the possible future introduction of emissions trading schemes, reductions in power and emissions might be achieved with design changes and fuel saving devices which are not necessarily the best economic solution.

6.6 Design procedures

The design process should be adapted to take account of the changing emphasis between economic viability and environmental factors, such as greenhouse gas emissions. The process will include some economic objective function, such as NPV or RFR, and an environmental objective function which could be the EEDI. A multiple criteria approach will be necessary, with weightings between the criteria depending on what financial incentives might arise in order to persuade ship operators to reduce emissions.

7. REFERENCES

1. MOLLAND, A.F. and UTAMA, I.K.A.P. Ship Wash Wave Analysis in Shallow Restricted Waters. *Third Regional Conference on Marine Technology, MARTEC'2002,* ITS Surabaya, Indonesia, July 2002.
2. MOLLAND, A.F., WILSON, P.A., TAUNTON, D.J., CHANDRAPRABHA, S. and GHANI, P.A. Resistance and Wash Measurements on a Series of High Speed Displacement Monohull and Catamaran Forms in Shallow Water. *Transactions of the Royal Institution of Naval Architects,* Vol.146, 2004.
3. CANDRIES, M. and ATLAR, M. On the drag and roughness characteristics of antifoulings. *Transactions of the Royal Institution of Naval Architects,* Vol.145, 2003.
4. IMO. Guidelines on the Method of Calculation of the Attained Energy Efficiency Design Index (EEDI) for new ships. Resolution of the Marine Protection Environment Committee, MEPC.212(63), 2012.
5. IMO. Guidelines on Survey and Certification of the Energy Efficiency Design Index (EEDI). Resolution of the Marine Protection Environment Committee, MEPC.214(63), 2012.
6. WOODYARD, D.F. *Pounder's Marine Diesel Engines and Gas Turbines.* 8th Edition. Butterworth-Heinemann, Oxford, UK, 2004.

7. MOLLAND, A.F. (ed) *The Maritime Engineering Reference Book.* Butterworth-Heinemann, Oxford, UK, 2008.
8. BSRA Methodical Series Experiments on Single-Screw Ocean-Going Merchant Ship Forms. Extended and Revised Overall Analysis. *BSRA Report* NS333, 1971.
9. KRACHT, A.M. Design of bulbous bows. *Transactions of the Society of Naval Architects and Marine Engineers*, Vol. 86, 1978.
10. HOCHKIRCH, K. and BERTRAM, V. Slow steaming bulbous bow optimisation for a large container ship. *Eighth International Conference on Computer and IT Applications in the Maritime Industries, COMPIT'09*, Budapest, 2009.
11. ANON The SHI SAVER Fin. Marine Power and Propulsion Supplement, *The Naval Architect*, RINA, London, 2008.
12. TOWNSIN, R.L., BYRNE, D., MILNE, A. and SVENSEN, T. E. Speed, power and roughness: The economics of outer bottom maintenance. *Transactions of the Royal Institution of Naval Architects,* Vol.122, 1980.
13. TOWNSIN, R.L., BYRNE, D., SVENSEN, T. E. and MILNE, A. Estimating the technical and economic penalties of hull and propeller roughness. *Transactions of the Society of Naval Architects and Marine Engineers*, 1981.
14. TOWNSIN, R.L., BYRNE, D., SVENSEN, T. E. and MILNE, A. Fuel economy due to improvements in ship hull roughness 1976-1986. *International Shipbuilding Progress*, Vol.33, 1986.
15. MOLLAND, A.F and TURNOCK, S.R. *Marine Rudders and Control Surfaces.* Butterworth-Heinemann, Oxford, UK. 2007.
16. ISHERWOOD, R.M. Wind Resistance of Merchant Ships. *Transactions of the Royal Institution of Naval Architects*, Vol. 115, 1973.
17. BERLEKOM, Van W.B. Wind Forces on Modern Ship Forms – Effects on Performance. *Transactions of the North East Institute of Engineers and Shipbuilders,* Vol. 97, No. 4, 1981.
18. GOULD, R.W.F. The Estimation of Wind Loadings on Ship Superstructures. *R.I.N.A. Marine Technology Monograph* No. 8, 1982.
19. MOLLAND, A.F. and BARBEAU, T-E. An Investigation into the Aerodynamic Drag on the Superstructures of Fast Catamarans. *Transactions of the Royal Institution of Naval Architects,* Vol.145, 2003.
20. DYNE, G. The efficiency of a propeller in uniform flow. *Transactions of the Royal Institution of Naval Architects*, Vol.136, 1994.
21. DYNE, G. The principles of propulsion optimisation. *Transactions of the Royal Institution of Naval Architects*, Vol.137, 1995
22. MOLLAND, A.F., TURNOCK, S.R. and HUDSON,D.A. *Ship Resistance and Propulsion.* Cambridge University Press, Cambridge, UK, 2011.
23. CARLTON, J.S. *Marine Propellers and Propulsion.* Second Edition. Butterworth-Heinemann, Oxford, UK, 2007.
24. WINTERS, R.D.E. Application of a large propeller to a container ship with keel drag. *Transactions of the Royal Institution of Naval Architects*, Vol.140, 1998.
25. SEO, K.C., ATLAR, M., CHUN, H.H. and KIM, H.J. The inclined keel concept – resistance analysis. *International Marine Design Conference, IMDC'09*, Trondheim, 2009.
26. ANON (2008d) The Mewis Duct. Marine Power and Propulsion Supplement, *The Naval Architect*, RINA, London, 2008.
27. ANON (2008e) The integrated propulsion manoeuvring system. *Ship and Boat International*, RINA, London, September/October 2008.
28. ECSA. Climatic change and shipping: ECSA position paper. European Community Shipowners' Associations (ECSA), January, 2008.
29. CARLTON, J., SMART, B. and JENKINS, V. The nuclear propulsion of merchant ships: aspect of engineering, science and technology. *Journal of Marine Engineering and Technology*, 2011.
30. LACKENBY, H. and PARKER, M.N. The BSRA Methodical Series – An overall Presentation: Variation of resistance with breadth-draught ratio and length-displacement ratio. *Transactions of the Royal Institution of Naval Architects*, Vol. 108, 1966.
31. HANSEN, H. and FREUND, M. Assistance tools for operational efficiency. *Ninth International Conference on Computer and IT Applications in the Maritime Industries, COMPIT'10*, Gubbio, 2010.
32. SATCHWELL, C. J. Windship technology and its application to motor ships. *Transactions of the Royal Institution of Naval Architects,* Vol.131, 1989.
33. RINA 1980 Proceedings of the Symposium on Wind Propulsion of Commercial Ships. *The Royal Institution of Naval Architects*, London, 1980.
34. WINDTECH'85 International Symposium on Windship Technology. University of Southampton, UK, 1985.
35. ANON. Christening and launch of 'E-Ship1' in Kiel. *The Naval Architect*, RINA, London, September, 2008.
36. ANON. Skysails hails latest data. *The Naval Architect*, RINA, London, September, 2008.
37. NYK. Eco Ship: Operation tests getting under way. NYK News Release, August 2004.
38. MOLLAND, A.F., TURNOCK, S.R. and HUDSON, D.A. Design metrics for evaluating the propulsive efficiency of future ships. *International Marine Design Conference, IMDC'09,* Trondheim, 2009.
39. SSPA HIGHLIGHTS. Newsletter published by SSPA, 3/2007.

40. SCHNEEKLUTH, H. and BERTRAM, V. *Ship Design for Efficiency and Economy*. 2nd edition, Butterworth-Heinemann, Oxford, UK. 1998.
41. SEN, P. Marine design: The multiple criteria approach. *Transactions of the Royal Institution of Naval Architects,* Vol.134, 1992.

8. AUTHORS' BIOGRAPHIES

I Ketut Aria Pria Utama is Professor of Naval Architecture at ITS, Surabaya, Indonesia. He received his MSc in 1996 and PhD in 1999 from the University of Southampton UK. He has extensive research in ship design and hydrodynamics. He is currently head of Department and Secretary of RINA Indonesian Branch.

Anthony Molland is Emeritus Professor of Ship Design at the University of Southampton, UK. His main areas of research are ship design, ship propulsion and rudders, and he has published widely on these subjects.

DESIGN ASPECTS OF OFFSHORE PLATFORM FOR COMBINING OF MARINE CURRENT AND OFFSHORE WIND ENERGY CONVERSION PLANT IN INDONESIA

A Suroso and **Murdjito**, Dept of Ocean Engineering-ITS, Indonesia

SUMMARY

Indonesia as a maritime country, its territory consists of 2/3 part of ocean waters and over than 17.000 islands spreading all the country. The waters consist of huge ocean energy which can be converted into others useful energy. The government had developed of a policy of energy diversification for increasing of the use of energy alternative. The policy is to anticipate the depletion of fuel energy where the oil resources estimated will be depleted around 20 years more. The authors selected marine current and offshore wind as the alternatives energy from the ocean.

The paper begins to introduce the potential site of the marine current throughout the Indonesia sea waters. Then the paper reviews of the development of marine energy conversion system. Selecting the type of offshore platform and the marine turbine is given to choose the type which is applicable in the potential site. Having selected the type, the paper is trying to discusses for combining of two types of energies (marine current and offshore wind) on one unit of platform. The design of the platform supporting of two energies is presented to end the paper. The paper was entirely constructed based on literature study

1. INTRODUCTION

The Indonesian government had developed of policy of energy diversification for increasing of the use energy alternative. The policy is to anticipate the depletion of fuel energy. The oil resources estimated will be depleted around 20 years more.

Indonesia as a maritime country, its territory consists of 2/3 part of ocean waters and over than 17.000 islands spreading all the country. Thousand of the Island have not been supplying of the electricity. The problem of supplying electricity for Small Island is the nature of the island. The location is .spreading, far away it's other. Therefore, the distribution of electrical energy would be a problem if large station of the energy located in one of the Island. In other word, small individual power station should been used such as Diesel generator.

As mentioned early, the fuel oil will be depleted in the near future. Therefore, study of alternative energies must be extended. The alternative energy such as solar and wind energy have been already developed. The solar energy is small capacity, it's only for lighting and TV. While, wind energy is not potential for certain small island due to Indonesia is geographically laid in equator area.

This paper is trying to introduce the energy from the sea i.e. marine current energy and offshore wind

2. POTENTIAL SITE OF MARINE CURRENT

As an archipelago country, Indonesia has many island and separate with sea and strait .where the current flows within it. As seen in the figure 1. Strong current exists from Pacific Ocean to Indian Ocean via Makasar and Lombok Strait. The current is also turn to the right and spread to the eastern of Indonesia.

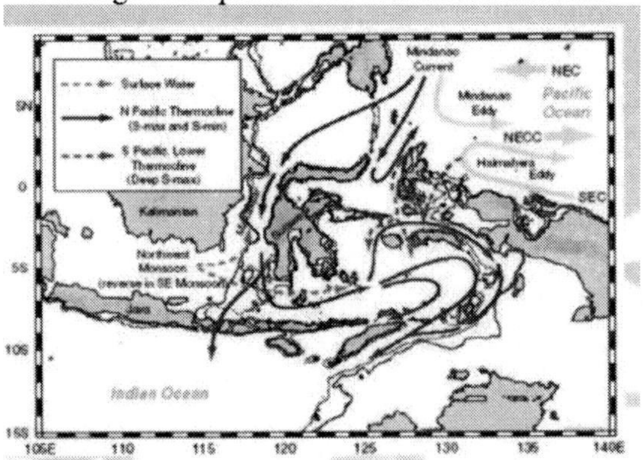

Figure 1. Indonesian Marine Current Map (Source; Kobolt Nusa)

According to mathematical model that energy per second intercept device of frontal area A (m^2) in water density of ρ, current velocity **V** (m/s) given by:

$$P=0.5\,\rho\,A\,V^3\,(t)$$

The formula indicates that the current speed should be greater than 1 m/sec due to the cubic speed.

Tidal currents are not constant. Generally they are a combination of quasi-steady current and flow induced by the tides. Estimation of energy captures therefore fairly complex procedure.

3. THE DEVELOPMENT OF MARINE CURRENT ENERGY CONVERSION PLANT

Effort to exploit of marine current energy has been developed by Marine Current Turbine Limited U.K.

This development is fully supported by the government. Nevertheless, the electricity has not been commercial product; the project has developed a proven device with the capacity of 300 KW. Hereby the name is Sea flow 300 KW. (figure 2) Full scale experiment was conducted during over than two years, in the offshore of Lynmouth, Devon. The experiment was very success. The system consists of an axial flow rotor with two blades of 20 meter of diameter The rotor drives a generator through a gearbox, similar to the wind turbine system. This company is looking for an industrial partner to be a commercial development.

Davis Hydro Turbine Limited installed marine current energy conversion near offshore British Colombia. A prototype of 1000 kW is tested since 1980. A unit of current energy conversion of 500 kW was also developed by the University of Edinburgh, U.K. Those development used vertical axis turbine due to the generator engine can be placed above sea surface. Further advantages are easy for operating and maintenance

Figure.2. Horizontal Turbine (source: MCT Ltd)

Gulf Stream of Florida offshore USA has been researched and planned to be deployed several turbines of 40 feet in diameter. Floating structures support the turbines in order easy to position the platform in the stream. The installation use turbines patented by Gorlow.

The current energy conversion is also developed in offshore South of Korea, where the installation consists of tenth of turbines which produced of 100 MW. Italian scientist selected the vertical axis turbine for their developing of marine current energy. They developed its own turbine called Kobold. Therefore the system called Kobold turbine and has been installed in the strait of Messina

From the development, there are two systems of turbine. They are Horizontal axis turbine and vertical axis turbine

Figure 3. Vertical Turbine
(Source: The Enermar System)

4. TURBINE BLADE

There are many turbine blades can be attached in the marine current power plant, such as NACA (Nautical Advisory Committee for Aeronautic), Kobolt Turbine, Gorlow turbine etc. The author created own design of the blade. The blade is adopted from the cross section of typical fresh ship called TOMBRO (see figure 4).
Compared to other fish, Tombro fish has better maneuvering. Based on the cross section we can design the blade.

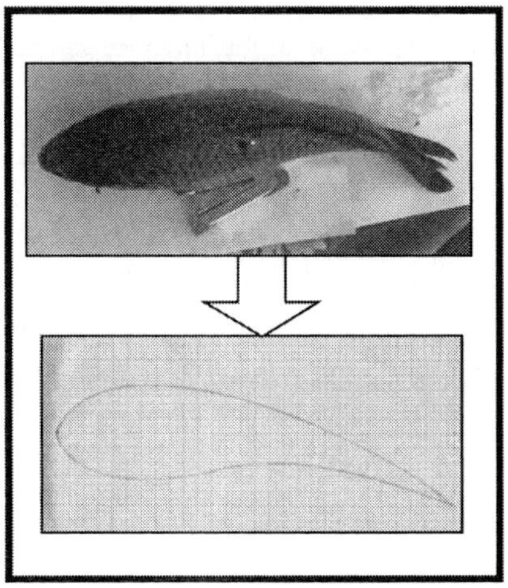

Figure 4. Blade Design

The design blade then to be calculated and analyzed its lift force by numerical method. The performing of the blade can be analyzed by hydraulic model test in wave tank.

5. BATHYMETRY FEATURE

The sea where the marine current plant installed has to be enough depth. The depth must accommodate the height of the platform and the water turbine. The platform can be fix or floating structure. Floating platform is more suitable choose due to of high tidal range in Indonesia. The site is also not too far from the coast line considering the high price and costly of marine cable. It can be said that the location of the plant is likely better in the steeply coastal.

6. BUILDING METHOD

The structure is fabricated as simple as possible in order that small fabrication yard with limited facilities can do the fabrication. Fabrication of the structure commences with keel laying as normally in the shipbuilding. The keel block is prepared prior to the keel laying. The fabrication grows up until the deck house and the structure is ready to be launched. The air bag is used as recently popular launching method. The advantage new technique is no slipway necessitated. The attachment of the turbine can be carried out when the structure has been floated and moored in some distant from the coast line. The turbine can be attached from the underwater of the structure by the helping of divers. Compared to the current turbine wind turbine is easier to attach on the top deck.

7. STRUCTURAL DESIGN

The structure is designed as like as ship designed, there are hull and deck house. The moon pool is placed in the centre of the hull to accommodate the shaft of the current turbine. The watertight tube is not necessitated to avoid the complicated of the structure and high cost. The connection between the shaft and machinery can be done in dry atmosphere. The advantage of this design is for the simplicity of the structural fabrication. The authors propose of marine hybrid energy of offshore wind and marine current plant, namely Windcur Energy (figure 5) The wind turbine selected is a vertical type due to its advantages such as simple tower and the delivers mechanical power at lower level which exhibits a low centre of gravity. Low centre of gravity improves the stability of the structure.

Figure 5. Concept of Windcur Energy

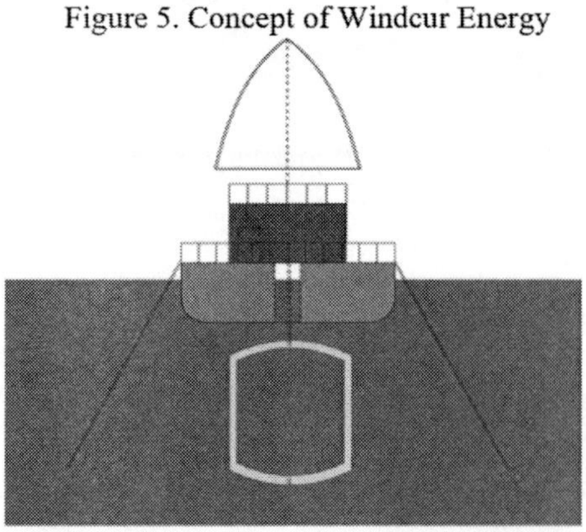

8. CONCLUSION

Having describing and discussion about the hybrid ocean energy power plant, some of conclusion can be highlight as follow:

1. Indonesia has many strong marine current for marine current energy plant
2. The best location is in the steeply coast
3. Propose of marine hybrid energy of offshore wind and marine current have been presented. However, the proposal design is very preliminary step. The design still need further study and research to become a proven technique.

ACKNOWLEDGEMENT

The authors wish to thank to the Department of high education to finance the research of Marine Current Energy

REFERENCES

Blue Energy at:http://www.bluenergy.com

Constans,J.,'Marine Sources of Energy', Pergamon Press, New York(1971)

Exim by Seapower,Delta and Stromturbine at:http://www.seapower.se/

Gorlov,A., 'Helical Turbines for the Gulf Stream', Marine Technology, Vol 35, No.3, July 1988,ppp 175-182 United State ,(1998)

MCT (Marine Current Turbine Ltd);htpp://www.marineturbine.com (2000)

Moroso,A and Eriksson,H.'The Enermar System', Technical Papers, Unido, Vienna (2005)

Suroso, A.' The Design of Marine Current Energy Conversion for Small Island Electricity',(in Indonesia) Research Report, Department of Higher Education, Jakarta(2007)

AUTHOR'S BIOGRAPHY

Arief Suroso is lecturer of Marine Energy Conversion System in the Department of Ocean Engineering-ITS. He has carried out some of research work of Marine Energy System funding by the high education department.

PRELIMINARY STUDY INTO THE SELECTION OF PASSENGER AND CARGO VESSELS FOR EASTERN INDONESIA

R. B Luhulima, Pattimura University, Ambon, Indonesia
I K A P Utama, Institut Teknologi Sepuluh Nopember (ITS), Surabaya, Indonesia
J-H Chen, National Cheng-Kung University (NCKU), Tainan, Taiwan

SUMMARY

The review and evaluation on the use of mono- and multi-hull (catamaran and trimaran) vessels for transporting passengers and cargoes for Maluku seawaters of Eastern Indonesia is carried out. The current study is unique because Maluku islands is typical archipelagic province of Indonesia where several islands are close together and some other islands are separated by oceans with high possibility of dangerous waves to occur hence can harm the safety of those vessels. The investigation is focused on the review and evaluation of resistance and powering, the choice of hull's material, seakeeping analysis, and the economic viability. The results are very useful in order to provide suitable, efficiient, and environmentally friendly vessels for that region.

NOMENCLATURE

Demihull	One of the hulls which make up the catamaran
L, L_{WL}	Length on waterline (m)
b	Breadth of demihull (m)
B	Breadth of catamaran (m)
T	Draught (m)
S/L	Lateral separation (clearance) ratio, between demihull centrelines
V	Ship Speed (m sec^{-1})
Fr	Froude number
Re	Reynolds number
R_T	Total resistance (N)
Cb	Block coefficient
C_T	Total resistance coefficient
WSA	Wetted surface area (static condition) (m^2)
υ	Kinematic viscosity (N s m^{-2})
ρ	Density of water (kg m^{-3})

1. INTRODUCTION

There is a significant development into the application of (especially) multihull vessels as ferries and other passenger and general cargo carriers in the last 30 years [1, 2]. The principal benefits of these vessel types compared to monohull one are those the multihull vessels are more attractive in term of layout accommodation, having better transverse stability with possibility to be as comfortable as monohull ships at certain sea-state, and in certain case this could reduce the total ship resistance and hence the size of main engine as well as the fuel consumption [3].

Various types of vessels have been developed to satisfy the design criteria and among others, the concept of catamaran is preferred and becoming more popular [4]. A number of works into the development of passenger vessels have been carried out such as done by Pal and Doctors [5] whom developed a preliminary design method to provide accurate solution of catamaran passenger vessel operated in a river. Meanwhile, Murdijanto et al [6] investigated the resistance and seakeeping characteristics of catamaran and trimaran for river and coastal operation at Indonesian seawaters.

2. RESISTANCE AND POWERING

Most of passenger and cargo vessels are of displacement monohull. This choice of the monohull is perhaps due to its advantages such as ability to cut through heavy waves easily, provide high loading capacity, good stability and seakeeping characteristics. For example, a passenger liner monohull can have good stability quality (satisfying the IMO criteria) and seakeeping characteristics [6, 7].

The resistance of monohull can be estimated using Equation (1).

$$R_T = 0.5\rho C_T (WSA) V^2 \tag{1}$$

The total resistance can be estimated using towing tank experiment, theoretical approach, empirical formulation such as developed by Holtrop and Mennen [8] and computational fluid dynamics (CFD) [9].

The area of hull under water, popularly known as WSA, becomes very important in order to optimise ship resistance and hence its power or the size of main engine. Large monohull such as tankers and giant passenger vessels tend to have high resistance and hence the power of engine. The power of main engine can be calculated by use of Equation (2)

$$P_E = R_T V \tag{2}$$

Higher resistance and hence higher power of engine has inspired naval architects to develop vessel to reduce wetted surface area. Static and dynamic stability requirement are later taken into consideration. Among others, multihull concept (catamaran and trimaran) has received considerable attention.

Catamaran, often called as 'cats' is twin-hulled watercraft, is noted for their speed and stability in open water. This type of vessel becomes very popular because having better transverse stability and providing wider deck area hence allow more freedom to ship designers in to construct layout arrangement [10].

The calculation of power required by the catamarans needs an investigation into the resistance characteristics entirely in order to obtain the most by ship design [11]. The resistance of catamaran can provide complex phenomena to ship designers particularly with the appearance of interaction between the demihull of catamaran. Therefore, it has been a basic need to obtain the breakdown and understanding of correct ship resistance components in order to obtain accurate calculation based on scaling transformation from model to the real ship.

A systematic investigation has been made by many researchers worldwide [10, 12, 13] showing that there is a certain separation between 2 demihulls hence causes interaction between the hulls. The interaction decreases as the separation to length ratio (S/L) increases. Wider separation such as S/L equals 0.4 and 0.5 indicate less interference between the hulls. This then provides an idea that a catamaran with similar displacement comparable to monohull could have smaller resistance and power of main engine.

Formulations to estimate the resistance of catamaran have been proposed by many researchers. However, among others, proposal by Insel and Molland [10] is found to be very systematic and then widely used to estimate the catamaran resistance and its resistance interference components. The complete formulation is written as Equation (3):

$$(C_T)_{CAT} = (1+\phi k)\sigma C_F + \tau C_W \qquad (3)$$

The factor ø has been introduced to take account of the pressure-field change around the demihulls and σ takes account of the velocity augmentation between the hulls and would be calculated from an integration of the local frictional resistance over the wetted surface, while (1+k) is the form factor for the demihull in isolation. For practical purposes, *ø* and σ can be combined into a viscous interference factor β where $(1+\o k)\ \sigma = (1+\beta k)$. Hence

$$(C_T)_{CAT} = (1+\beta k)C_F + \tau C_W \qquad (4)$$

Further investigation on the catamaran resistance is pioneered by Soeding [14] who found out that the reduction of ship resistance significantly when the demihull is varied longitudinally and this is known as staggered catamaran. Jamaluddin et al [15, 16] investigated various hull configuration both symmetric and asymmetric hulls, found out that the reduction of resistance occurs when the catamaran was varied transversely (un-staggered) and longitudinally (staggered). Murdjanto et al [6], in particular, studied the possibility of river catamaran to carry passengers. It was found out that the resistance of catamaran and hence its power could be less than the power of monohull with equal displacement. If this is applied to a real ship, it has the potency to reduce the cost of engine and safe the use of fuel significantly.

The trimaran hull form, which is also known as vessel with three hulls, has received considerable attention because it can provide wider deck area and shallower draft hence safer for shallow draft operation [6, 17]. The form of trimaran is recognised at the beginning as 'ship with outriggers' and at the present time is popularly used as warships and passenger carriers attributed to its high quality of manoeuvring and stability [18].

3. THE CHOICE OF MATERIAL

There are a number of choices of material to build a vessel. Some of those include aluminium, steel, composite or combination of them. In term of strength, steel can be very strong but it may be too heavy hence become less suitable particularly for catamaran and trimaran. Aluminium is also appropriate because having significant strength and lighter than steel, but it is probably more expensive than steel.

Fibreglass reinforced plastics (FRP) or any kind of composites could be another alternative as material for ship construction. It is even much lighter than steel and having significant strength to resist certain loading conditions. However, the process of composing FRP may face environmental problems because it involves the use of chemical (and toxic) products. Examples of proposed method for material selection were given in [19, 20].

4. SEAKEEPING EVALUATION

Performance of ship at sea usually known as seakeeping and this is among the most important factors when comparing competing vessels or vessel types. This is due to the seakeeping criteria, which can influence important aspects such as passenger comfort, operational limits, speed loss and structural integrity [21].

Motion transfer functions for each vessel and wave energy spectra for the relevant sea area are necessary to perform spectral calculations. Kurniawan [22] provides good information about wave spectrum of Indonesian waters, including Moluccas seawaters. This allows the calculation of statistical quantities or attributes such as the RMS values of the various motions and accelerations and the probabilities of an individual motion or acceleration exceeding a given value. This also assists ship designers to predict vessel's seakeeping characteristics more realistic.

Example of seakeeping calculation based on experimental investigation was given in [6]. Three types of vessel were tested, namely monohull, catamaran and trimaran. The tests were conducted as sea state of 3 and foud out that the multihull modes have better seakeeping characteristics than the monohull, although the discrepancy is not significant. In addition, among the multihulls, the catamaran showed slightly smaller heave and pitch. In contrast, trimaran demonstrated smaller roll response than catamaran. The reason for this is due to the number of hulls of trimaran and hence the total ship breadth.

5. ECONOMIC VIABILITY

Economic evaluation is aimed to make sure the vessel's potential to make a profit during its service life [23]. Consequently, the proposed vessels should be compared directly against each other, in order to determine of the proposal is an appropriate investment for the clients.

The economic viability evaluation, however, is limited to the following aspects such as the choice of vessel types (monohull or multihulls), the choice of material (steel, aluminium or composite) and the choice of powering system (conventional engine, sail, electric or combination). There is significant trend currently worldwide to reduce the use of conventional engine along with fossil fuel. The reason for this is attributed to the rare stock as well as the continuous increase price of that fuel.

The complete replacement of engine and fossil fuel seems to be impossible at present. The use of (say) hybrid electric and wave power mechanism cannot satisfy the service speed usually obtained by using the current system. It does affect the delivery time of transporting cargoes and passengers hence this choice is likely not economic.

There is also problem considering the increased pressure from the environmental viewpoint and with the possible future introduction of emissions trading schemes, reductions in power and emissions might be achieved with design changes and fuel saving devices which are not necessarily the best economic solution [24].

6. CONCLUSIONS

There are several choices on the use of vessels to carry out passengers and cargoes, either monohull or multihulls. Monohull type may be more comfortable than multihulls, but it might be not efficient (say) at higher speed condition. Multihulls may be more stable but may be more difficult to handle.

There are a number of material choices such as using steel, aluminium and FRP. Each has advantages and disadvantages. Steel may be the most strong but it is too heavy if used for high speed craft. Meanwhile, FRP is lighter and strong enough, but facing serious environmental problem.

Seakeeping criteria is very important in correlation with comfortability. Both monohull and multihull can have equal dynamic motions (heave, pitch and roll) especially at low sea state (say 3) hence those vessels are very comfortable to operate. In addition, the availability of local sea spectra helps ship designer a lot in order to evaluate the vessel's seakeeping characteristics better.

The use of alternative power sources to replace the use of conventional engine along with fossil fuel is believed to be very important. However, there is no apparent data supporting that expectation. There is worrying about the economic consequences of the proposed replacement which is found to be not the best economic solution.

7. REFERENCES

1. UTAMA, I K A P, *PhD Thesis*, Department of Ship Science, University of Southampton, UK, 1999.
2. ITAMA, I K A P and MOLLAND, A F, *FAST 2001*, Southampton UK, 3-5 September 2001.
3. FRY, E D and GRAUL, T (1972), Design and Application of Modern High-Speed Catamarans, *Marine Technology*, Vol. 9, No. 3, pp 345-357.
4. SAHOO, P.K., SALAS, M and SCHWETZ, A, *Ships and Offshore Structures*, Vol. 2, No. 4, 2007.
5. PAL, P K and DOCTORS, L J, Proceedings FAST '95, Lubeck-Travermunde, Germany, 1995.
6. MURDIJANTO, UTAMA, I K A P and JAMALUDDIN, A, An Investigation into the Resistance/Powering and Seakeeping Characteristics of River Catamaran and Trimaran, *Makara Seri Teknologi*, Vol. 15, No. 1, 2011.
7. LEWANDOWSKI, E M, The Dynamics of Marine Craft: Maneuvering and Seakeeping, *Advanced Series on Ocean Engineering*, Vol. 22, World Scientific, Singapore, 2004.
8. HOLTROP, J and MENNEN, G G J,
9. ZHANG Z, LIU H, ZHU S, ZHAO F., Application of CFD in Ship Engineering Design Practice and Ship Hydrodynamics, *Conference of Global Chinese Scholars on Hydrodynamics*, 2006.
10. M and MOLLAND, A F, An Investigation into the Resistance Components of High Speen Displacement Catamaran, *RINA Transactions*, Vol. 134, 1992.
11. MOLLAND, A.F. (2008), *A Guide to Ship Design, Construction and Operation*, The Maritime Engineering Reference Book, Butterworth-Heinemann, Elsevier.

12. COUSER, P R, MOLLAND, A F, ARMSTRONG, N and UTAMA, I K A P, Calm Water Powering Predictions for High Speed Catamarans, *Procs of FAST'97*, Sydney, Australia, July 21-23, 1997.
13. MOLLAND, A F, UTAMA, I K A P and BUCKLAND, D, Power Estimation for High Speed Displacement Catamarans, *Procs MARTEC 2000*, Surabaya, Indonesia, September 7-8, 2000.
14. SOEDING, H, Drastic Resistance Reductions in Catamarans by Staggered Hulls, *Procs of FAST'97*, Sydney, Australia, July 21-23, 1997.
15. JAMALUDDIN, A, UTAMA, I K A P, ARYAWAN, W D and WIDODO, B, Experimental Investigations into the Resistance Components of Symmetrical Catamarans with Variations in Hull Clearances and Staggers, *RINA Transactions*, Vol 154, Part B1, 2012.
16. JAMALUDDIN, A, UTAMA, I K A P, ARYAWAN, W D and WIDODO, B, Experimental Investigations into the Resistance Components of Asymmetrical Catamarans with Variation in Hull Spacing and Stagger, *RINA Transactions*, Vol 154, Part B1, 2012.
17. SUBRAMANIAN, V A, DHINESH, G and DEEPTI, J M, *the Journal of Ocean Technology*, Canada's Arctic, Vol. 1, No. 1, 2006.
18. HEBBLEWHITE, K, SAHOO, P K and DOCTOR L J, *Ships and Offshore Structures*, Vol. 2, No. 2, 2008.
19. MOLLAND, A F, KARAYANNIS, T and COUSER, P R, Concept Exploration and Assessment of Alternative High Speed Ferry Types, *Procs FAST'97*, Sydney, Australia, July 21-23, 1997.
20. KARAYANNIS, T and MOLLAND, A F, Selection between Alternative High Speed Ferries Based on Design Robustness, *Procs HIPER*, 1999.
21. RAWSON, K J and TUPPER, E C, *Basic Ship Theory*, Vol. 2, Longman Scientific and Technical, Harlow, UK, 1993.
22. KURNIAWAN, A, *MSc Thesis*, Faculty of Marine Technology, ITS, Surabaya, 2012.
23. THOMAS, G, TOMIC, P and TUITE, High-Speed Catamaran or Monohull? How do you choose? *5th International Conference on High Performance Marine Vehicles*, Australia, 8-10 November, 2006.
24. UTAMA, I K A P and MOLLAND, A F, The Powering of Future Ships taking into Consideration Economic Viability and Environmental Issues, *Procs ICSOT Indonesia 2012*, Ambon, Indonesia, 7-8 November 2012.

8. AUTHORS BIOGRAPHY

Richard Beny Luhulima, MT is a lecturer at the Pattimura Univesity in Ambon, Indonesia. He is currently a PhD student at the Department of Naval Architecture and Shipbuilding Engineering under supervision of Prof Utama. His research focuses on the development of various hull configuration from mono- to multi-hulls as passenger and general cargo carriers for archipelagic waters such as Moluccas Islands at Eastern Indonesia.

I Ketut Aria Pria Utama, M.Sc, Ph.D, CEng, FRINA is a Professor of Ship Hydrodynamics at the Department of Naval Architecture and Shipbuilding Engineering, ITS Surabaya. Prof. Utama received his M.Sc in 1996 and Ph.D in 1999 from the University of Southampton, UK. He has extensive research areas in ship design and hydrodynamics. He is presently Head of Department and Secretary of the RINA Indonesian Branch.

Jeng-Horng Chen, MSc, PhD is an associate professor at the Department of Mechatronics and Marine System Engineering, National Cheng-Kung University (NCKU) at Tainan, Taiwan. He has extensive research in hydrodynamics and presently as chairman of NCKU maritime museum.

THE EFFECTS OF HULL MODIFICATION ON DESIGN PARAMETERS OF MEDIUM-SPEED MONOHULL PASSENGER FERRIES

W R Hetharia, A Hage and Ph Rigo, Université de Liége, Belgium

SUMMARY

The development of medium-speed monohull passenger ferries has shown their better performance in some maritime countries recently. With the operational speeds may reach 23 knots, those ships are classified as semi-displacement ships where the range of those speeds are beyond the hump speed. In fact, those ships need a lot of energy to maintain their speeds. Since the database was not available and provided for the optimization process of those ships then a base ship of 254 passengers was designed in this study. The modifications of hull dimensions of the base ship were executed due to the ship layouts. The effects of changes in resistance, stability and other design parameters were analyzed in this study. An example of optimization process for one configuration of ship layout is presented in this paper.

NOMENCLATURE

$A_1..A_{27}$	Coefficients in equation (1)
A_T	Transom area (m^2)
A_X	Maximum section area (m^2)
B	Molded beam of the submerged hull (m)
C_B	Block coefficient
C_M	Midship coefficient
C_P	Prismatic coefficient
C_{WP}	Waterplane coefficient
C_∇	Beam-loading coefficient
D	Molded depth (m)
Fn	Froude number (nondimensional)
GM_T	Transverse metasenter height (m)
L	Molded ship length (m)
L_{CB}	Longitudinal center of buoyancy (m. %L)
P_B	Brake power (kW)
P_E	Effective power (kW)
R_T	Total resistance (kN)
T	Design molded draft (m)
T_h	Heave natural period (s)
$T\phi$	Roll natural period (s)
T_Θ	Pitch natural period (s)
U	Coefficient in equation (1) = $\sqrt{2 i_e}$
V	Ship speed (m/s, knot)
W	Ratio A_T/A_X in equation (1)
X	Ratio $\nabla^{1/3}/L$ in equation (1)
Z	Ratio ∇/B^3 in equation (1)
w	Taylor wake fraction
t	Thrust deduction factor
α_E, i_e	Angle of entrance of load waterline (degree)
η_h	Hull efficiency
η_o	Propeller efficiency
η_r	Relative rotative efficiency
η_s	Seal efficiency
η_b	Line shaft bearing efficiency
∇	Molded volume at the design waterline (m^3)
Δ	Displacement at the design waterline (t)

1. INTRODUCTION

Recent development of medium-speed passenger ferries has brought a new challenge in the maritime fields. Those ships operate in some regions of the world such as Mediterranean, Asia, Pacific, Central America and Europe with the speed range of 18 to 25 knots. The existence of those ships is to fulfill an empty speed region between conventional ferries and high speed crafts (HSC). Since the emerging, those ships had been built and operated in all regions of the world for the specific routes. The application of the hull material of Aluminum in recent years to those ships gives the benefits for the additional payload or reducing the power. Those ships are operating in short-sea distance where most of them are multi-hulled type (catamarans). However, due to their simplicity, the monohulls type are developed also and have a potential future markets. In fact, those ships operate at the range of Froude numbers Fn from 0.55 to 0.80 which is beyond the hump speed ($Fn > 0.50$). Therefore, they need a great engine power to maintain their service speed. The efforts should be done in order to optimize the engine power of the ship. To minimize the engine power or ship resistance then the hull dimensions and geometrical forms should be taken into account. However this effort would affects other design parameters. From the existing database of ships, it was found that for an input of number of passengers those ships have a difference in dimensions and engine power. In this study, a base (parent) ship was designed and modified for its dimensions. The modification process is conducted based on the layout of the ship. The layout is arranged due to the passenger distribution along the ship length and ship beam. In addition, the layout is arranged also for passenger distribution on main deck and upper decks. This arrangement will end-up with the variation of ship length and beam. Furthermore, one configuration of those layouts is evaluated and optimized in this study.

2. LITERATURE REVIEW

The medium-speed passenger ferries are classified as semi-displacement ships [1, 2] and in other references it is known also as semi-planing [3]. From the structural point of view, due to their speeds and lighter displacements, they are classified as High Speed Craft (HSC) [4]. The term of medium-speed may be found in Reference [2]. The recent term for medium-speed may be found in the maritime fields where the speeds is not exceed 25 knots. Due to their operational range those ships are classified as short-sea ferries and have the speeds that do not generally exceed 25 knots [5].

2.1. THE EFFECTS OF DIMENSIONS AND FORMS FOR SHIP PERFORMANCE

The products of a ship design are dimensions and geometrical hull forms. These products, in fact, should satisfy the best performance of the ship. This may be achieved by modifying them in such a way until they fulfill the target of final result. Modification the dimensions and hull forms may be executed by changing them in some patterns. In fact, these may affect the design parameters of the ship. As stated by Watson [6] and Parsons [7] that the changing of hull dimensions (length, breadth, draft, height) and hull forms (coefficients) will affect the powering, cost, stability and other ship parameters. In fact the modified hull dimensions and forms may be applied further for the optimization process due to the required optimization objectives. Beside the ship dimensions, the hull forms play the important roles also for the ship parameters. They are represented by C_B, C_M, C_P, C_{WP}, L_{CB}, α_E. They may be found at some references for some types of ships. Also for optimization process, the range of values of those coefficient may be found in References [7,8].

2.2. RESISTANCE AND PROPULSION OF SEMI-DISPLACEMENT SHIPS

Due to their special hydrodynamic aspects, some studies has been executed concerned these ships. Molland [1] presents some statistical data for the computation of resistance of semi-displacement ships such as dimensional ratios of $L/\nabla^{1/3}$ = 6.0 to 9.0; L/B = 5.0 to 7.0; B/T = up to 5.0 which affect the resistance and stability. Larsson [3] presents the information concerning the best value of transom area (A_{tr}) for those semi-planning ships.

The systematic series of resistance data of semi-displacement ships that may be used as a basis for preliminary power estimates are found in References [1, 3, 9, 10]. From all the existing resistance series, the resistance of the semi-displacement ships depend on the parameters such as [1, 10]:

- Length displacement ratio = $L_{WL}/\nabla^{1/3}$
- Beam-loading coefficient $C_\nabla = \nabla/B^3$.
- Angle of entrance of the load waterline = i_e
- Ratio of transom area to maximum section area = A_T/A_X

Meanwhile, some series methods also include the parameters such as: ratio of B/T, LCB, C_B, C_P, C_{WP}.

Two statistical methods were suitable for the computation of ship resistance for those kind of ships, i.e. Savistsky pre-planing [9] and WUMTIA [1]. The statistical resistance prediction method derived by Mercier and Savitsky is suitable for the semi-planning ships. The general form of the resistance equation adopted by Mercier and Savitsky is as follows:

$$R_T/W = A_1 + A_2X + A_4U + A_5W + A_6XZ + A_7XU + A_8XW + A_9ZU + A_{10}ZW + A_{15}W^2 + A_{18}XW^2 + A_{19}ZX^2 + A_{24}UW^2 + A_{27}WU^2 \quad (1)$$

where: $X = \nabla^{1/3}/L$; $Z = \nabla/B^3$; $U = \sqrt{2i_E}$; $W = A_T/A_X$.

The values of the coefficients A_1 to A_{27} and correction factors are presented in Reference [9]. This method is provided in the Maxsurf software.

The effective power of the ship is computed as:

$$P_E = R_T \times V \quad (2)$$

where: R_T = total resistance and V = speed of the ship

The engine power (brake power P_B) is computed in relation with the effective power P_E [7].

$$P_B = P_E / (\eta_h \eta_o \eta_r \eta_s \eta_b \eta_t) \quad (3)$$

where:

η_h = hull efficiency;
η_o = propeller efficiency;
η_r = relative rotative efficiency = 1.0
η_s = seal efficiency;
η_b = line shaft bearing efficiency;
η_t = transmission efficiency;
$\eta_s \eta_b$ = 0.97 for machinery amidships
η_t = 0.975 for medium speed diesel plant

Hull efficiency is computed as:

$$\eta_h = (1 - t)/(1 - w) \quad (4)$$

where:

w = Taylor wake fraction = $0.5C_B - 0.05$ (5)
C_B = block coefficient
t = thrust deduction factor = $0.6\,w$ (6)

The maximum continuous rating (MCR) of the main engine is determined by adding a power service margin as 10% to the brake power.

$$MCR \geq (1 + MS)\, P_B \quad (7)$$

where: MS = power service margin

Two screw propeller units are used for the ship. The screw propellers are evaluated based on the propeller data from the Wageningen B-Screw Series [9]. The evaluation for the cavitations of the propellers is executed based on Burril Diagram of cavitations.

2.3. STABILITY REQUIREMENTS FOR MEDIUM-SPEED PASSENGER SHIPS

The stability criteria that should be fulfilled by all passenger ships is not only due to the initial metasenter height (GMTo). In fact, the ship stability criteria should fulfilled at the large angle of inclinations. The stability criteria that should be applied for the ship stability are based on the HSC Code 2000 MSC 97(73)-Annex 8 Monohull Intact, HSC Code 2000 Chapter 2 Part B Passenger Craft Intact and IMO MSC 36(63) HSC Code Monohull. Those criteria include:

- Angle steady hell ≤ 16 deg
- Angle steady hell/margin line immersion ≤ 80%
- Area1/Area2 ≥ 100 %
- Area 0-GZ_{max} ≥ 0.055 m.rad
- Area 30 to 40 ≥ 0.03 m.rad
- Angle GZ_{max} ≥ 15 degree
- Initial GMt ≥ 0.15 m
- Angle passenger crowd ≤10 degree
- Angle high speed turning ≤10 degree

It is quite difficult to find this information from the existing similar ships. Therefore, a parent ship is designed and provided the parametric model data for the future optimization process.

2.4. SEAKEEPING REQUIREMENTS FOR MEDIUM-SPEED PASSENGER SHIPS.

For the initial ship design, the sea keeping parameters were evaluated for rolling, pitching and heaving natural periods [7]. Those are computed due to the following formulas:

Roll natural period: $T\phi = 2.007\, k_{11} / \sqrt{GM_T}$ (8)

where: $k_{11} = 0.40\, B$

Pitch natural period:

$T_{\Theta} = 1.776\, C_{WP}^{-1} \sqrt{(T\, C_B\, (0.6 + 0.36\, B/T)}$ (9)

Heave natural period:

$T_h = 2.007 \sqrt{(T\, C_B (B/3T+1.2)/C_{WP})}$ (10)

3. DATABASE OF PARENT SHIP

Since it is quite difficult to find the database of similar ship with the hull material of Aluminum, then a base ship was designed in this study. Generally, the data of the existing ships are provided such as main dimensions, speed and engine power. Other detail data such as details of weights, centers and geometrical hull forms are quite difficult to be found. Stability is an important parameter that is required for medium speed passenger ferries. The criteria of stability at the large angle of inclination (*GZ* curve) are evaluated during the design process. One important objective of design the base ship is to avoid some uncertainty factor during the modification or optimization process [7]. If the database is not available, one way to do is designing the parent ship. The results of the parent ship would be explored next for a specific objective function. As stated by Gale [11]: "The point designs, once they have been developed, can be used as parents to explore the effects of parametric variations in other, second order parameter".

In the design process, a number of cycles (iterations) are required to arrive at a satisfactory solution [11, 6, 7]. Particularly in designing a passenger ferry several important key factors should be concerned and included [11, 12, 13, 14, 15]. Those key factors include: spaces, access, services rooms, arrangement, accommodation and safety for passengers, services and facilities for ship operations. Also scantling, weights and centers, stability, safety and seakeeping are important parameters for ship operation. All requirements and rules imposed for the ship design are strictly required during design process. The parameters of parent ship are presented in Table 1. The layout of the ship is shown in Fig. 1.

Parameters	Value	Unit
Total number of passengers	254	
Number of seats in row	10	
Number of crews	5	
Number of pax main-upper decks	177-77	
Pax distributions main-upper decks	70-30	%
Service speed	19.98	knots
Navigation range	200	n.m
Length overall, L_{OA}	32.00	m
Length of waterline, L_{WL} (= L_{BP})	29.00	m
Ship beam (at main deck), B	7.00	m
Ship beam (waterline), B_{WL}	6.69	m
Draft, T	1.388	m
Deck height, D	2.600	m
Displacement, Δ	105.3	Tone
Volume displacement ∇	102.7	m^3
Block coefficient, C_B	0.382	
Midship coefficient, C_M	0.547	
Prismatic coefficient, C_P	0.697	
Water plane coefficient, C_{WP}	0.76	
Long. C. of buoyancy, Lcb (fr AP)	48.33	% Lwl
Vertical center of buoyancy, KB	0.967	m
Radius metacenter, BM_T	5.374	m
Vertical center of gravity, KG	3.039	m
Initial metacenter, GM_T	3.302	m
Total resistance (at 19.98 knots)	93.6	kN
Total engine power (MCR)	2402	hp
Total propulsion efficiency η_t	0.588	
Period of rolling, Tϕ	2.97	Second
Period of pitching, T_{Θ}	2.31	Second
Period of heaving, Th	2.65	Second

Table 1: Parameters of the parent ship

Other ship database are described as follows:

- The structure components of the ship were determined based on the Rules for the Classification of High Speed Craft [4].
- Type of ship: passenger ferry/class B.
- Type of passenger accommodation: seat in passenger saloon.
- The hull material of the ship: Aluminum Alloy. The types of alloys used for the ship are 5083 H111 for plating and 6082 T6 for profile.

- Type of main engine: MTU Marine Diesel Engine 10V 2000 M70.
- Rated power: 1205 bhp (Rated speed is 2250 rpm).
- Gearbox: ZF 3000 (i = 2.0)
- Type of selected screw propeller: B 4-70 (BAR : 0.70, Number of blades: 4, P/D: 0.799, Diameter: 1.078 m, Efficiency: 0.584)
- Maximum speed for the ship: 19.98 knots

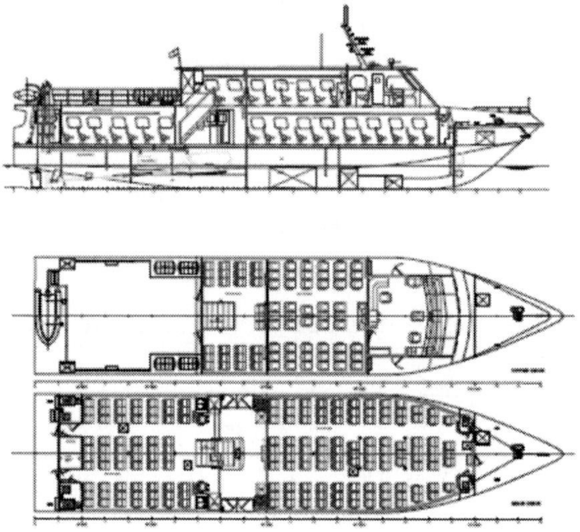

Figure 1: Layout of the parent ship

The total weight of the ship is computed directly from the parent ship. The weights and their centers are presented in Table 2. A margin weight of 4% is added to the ship lightweight. In addition, a margin of VCG of 0.150 m was added for the VCG of the ship lightweight [7].

Items weight	Weight (tone)	LCG (m)	VCG (m)
Structural weight	38.902	-0.628	2.882
Machinery and systems	15.725	-2.900	1.418
Ship outfits	13.960	-0.381	3.773
Sum LWT	68.587	-1.099	2.728
Margin weight	2.743	-1.099	2.728
DWT (pax & liquids)	33.922	0.764	3.355
Total weight	105.252	-0.498	2.930

Table 2: Weight and centers of the parent ship

4. MODIFICATION OF HULL DIMENSIONS

As stated earlier that the resistance of the semi-displacement ships depend on the parameters length displacement ratio $L_{WL}/\nabla^{1/3}$ and beam-loading coefficient $C_{\nabla} = \nabla/B^3$. In fact, for the small passenger ships such as medium-speed passenger ferries this may be achieved by the layout of the ship. For the real case, the layout is arranged by passenger distributions due to the length, width as well as main-upper decks. The parent ship has 254 passengers. Numbers of seats for passengers in a row due to the ship beam are 10 seats. The ship was named for P254S10. The passenger distribution main-upper deck is 70-30%. The similar way was done for the same number of passengers with number of seats are 9. The ship was named for P254S9. The distribution of passengers was done by shifting each seat row of passengers from main deck to upper deck or vice versa. It was noticed that during the shifting of seats, the weight components of passengers and their belongings were not changing. The only effect of movement of passengers was the changing of vertical and longitudinal centers of weights. In fact, this shifting of passengers affected the ship length and beam due to the required space. In addition, the weight of structural items and its centers were changed. As a result the total ship weight or displacement of the ship was changed.

During the modification for P254S10 and P254S9 due to the pax distribution main-upper decks, the main dimensions such as ship beam, deck, midship cross-section and transom area were kept to be constant. The main changing of the ship parameters were: beam, length, displacement, block and prismatic coefficients, longitudinal center of buoyancy, and half angle of entrance. The main difference of P254S10 and P254S9 is ship beam which is 7.00m for P250S10 and 6.50 m for P254S9.

Assume that during the modification of ship length, the structural elements were constants then structural weight was changed. To add or reduce one meter length of ship for P254S10, the longitudinal structural weight will change as much as 0.818 tone (2.10 % of total structural weight) for all structures below the main deck. With the addition of transverse structures this will change 1.033 tone (2.65 % of total structural weight). In addition, for the structure components of the upper deck, weight changes by 0.235 tone (0.6 %) for longitudinal structures and 0.262 ton (0.67 %) for longitudinal and transversal structures. The similar way for P254S9, the longitudinal structural weight will change as much as 0.775 tone (1.991 % of total structural weight) for all structures below the main deck. With the addition of transverse structures this will change 0.976 tone (2.51 % of total structural weight). In addition, for the structure components of the upper deck, weight changes by 0.228 tone (0.59 %) for longitudinal structures and 0.254 tone (0.65 %) for longitudinal and transversal structures. The structural weight decreases as much as 1.256 tone (3.23 %) as the ship configuration changed from P254S10 to P254S9. The ship parameters due to the modification of length and beam are presented in Table 3.

The parameters of ship resistance and stability were computed by using Maxsurf software. The criteria that have been used for the ship stability are based on HSC Code 2000 MSC 97(73) - Annex 8 Monohull Intact, HSC Code 2000 Chapter 2 Part B Passenger Craft Intact and IMO MSC 36(63) HSC Code Monohull Intact. In addition, the ship resistance was computed based on the statistical method of Savitsky pre-planning. The results of the computations are shown in Figures 2 to 9.

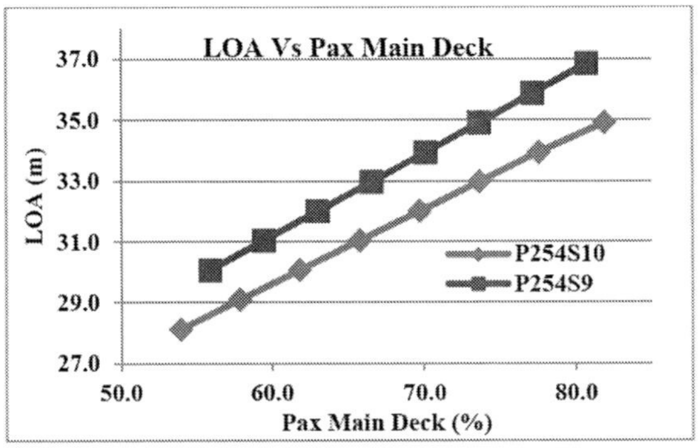

Figure 2: Ship length due to passenger distribution on main deck

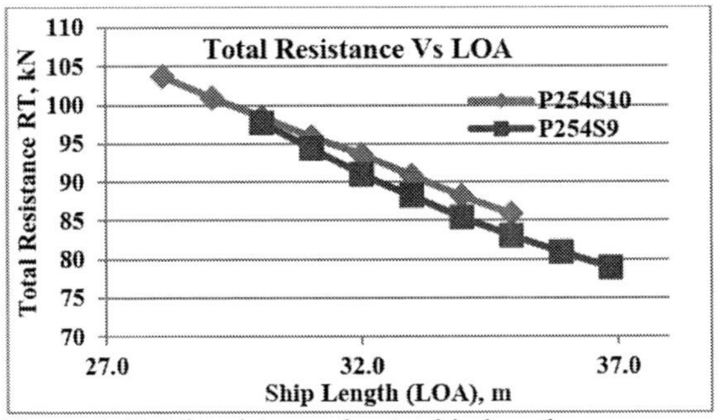

Figure 3: Total resistance due to ship length

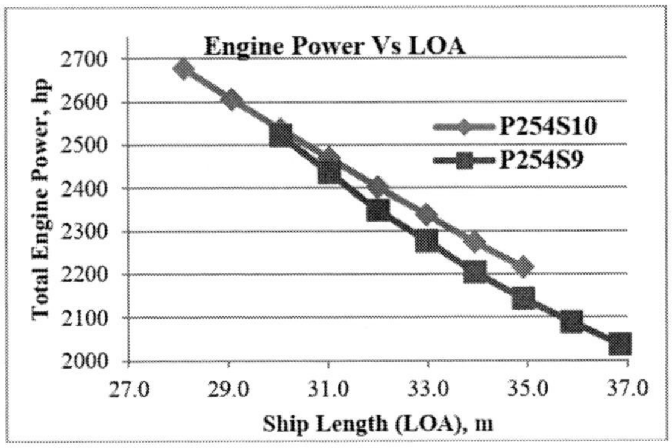

Figure 4: Total engine power due to ship length

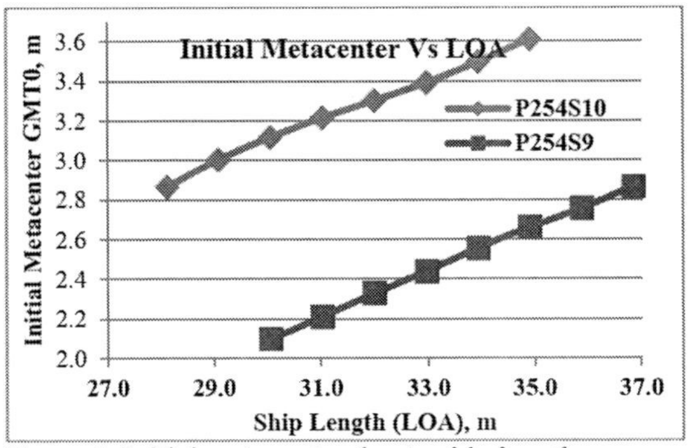

Figure 5. Initial metacenter due to ship length

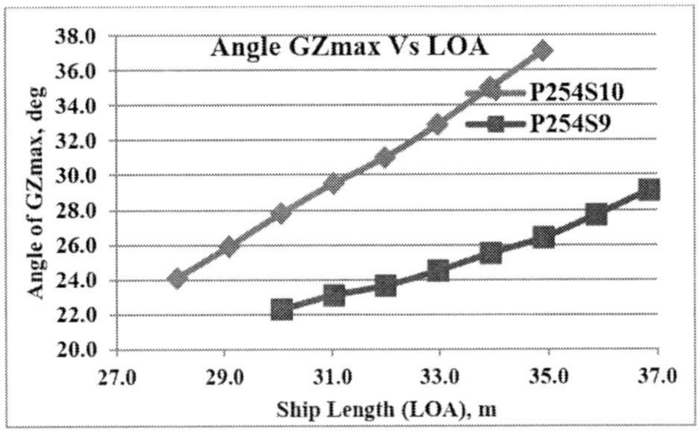

Figure 6. Angle of GZmax due to ship length

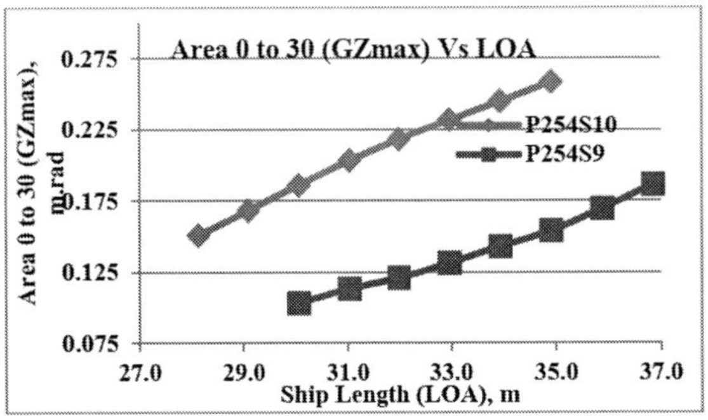

Figure 7. Area 0 to 30 due to ship length

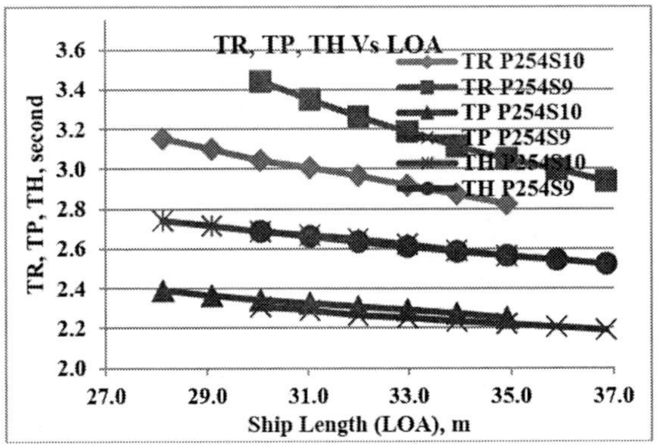

Figure 8. Periods of Tr, Tp and Th due to ship length

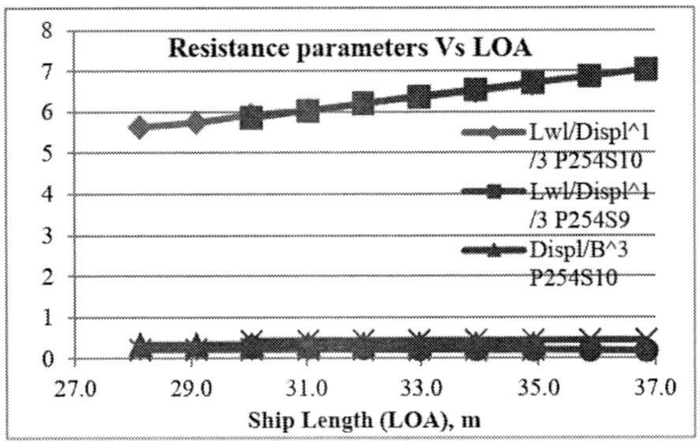

Figure 9. Resistance parameters due to ship length

It may be seen from the Figure 2 that the modification of ship layout due to seat arrangements (S9 and S10) and passenger distribution main-upper deck affect the ship length. The figures 3 and 4 show that as the length increase then the average total resistance or engine power decrease 2.7 % for P254S10 and 3.0 % for P254S9. Meanwhile, for the equal lengths of P254S10 and P254S9 the total resistance or power is less for P254S9. The average difference is 2.8 %.

From the Figure 5 it is seen that the initial metacenter increases as length increases. The average value of increasing initial metacenter is 3.6 % for P254S10 and 4.6 % for P254S9. Meanwhile, for the equal lengths of P254S10 and P254S9 initial metacenter is lower for P254S9. The average difference is 29.4 %.

From the Figure 6 it is seen that the angle of *GZ*max increases as length increases. The average value of increasing *GZ*max is 6.4 % for P254S10 and 3.9 % for P254S9. Meanwhile, for the equal lengths of P254S10 and P254S9 the angle of *GZ*max is lower for P254S9. The average difference is 23.7 %.

From the Figure 7 it is seen that the area 0 to 30 increases as length increases. The average value of increasing area 0 to 30 is 8.0 % for P254S10 and 8.8 % for P254S9. Meanwhile, for the equal lengths of P254S10 and P254S9 the area 0 to 30 is lower for P254S9. The average difference is 44.5 %.

From the Figure 8 it is seen that the rolling period decreases as length increases. The average value of decreasing rolling period is 1.6 % for P254S10 and 2.5 % for P254S9. Meanwhile, for the equal lengths of P254S10 and P254S9 the rolling period is higher for P254S9. The average difference is 10.1 %.
Figure 9 presents other resistance parameters such as length-displacement ratio $L/\nabla^{1/3}$ which are 5.63 to 6.68 for P254S10 and 5.86 to 7.03 for P254S9. Beam-loading ratios ∇/B^3 are 0.33 to 0.35 for P254S10 and 0.42 to 0.44 for P254S9. Ratios of area transom to area maximum A_T/A_M are 0.216 to 0.203 for P254S10 and 0.281 to 0.179 for P254S9.

5. OPTIMIZATION PROCCESS FOR P254S10

Since the modification of the ship was done due to the ship length, then all parameters of resistance, stability and seakeeping are presented as the function of ship length. From Figures 3 to 9 it seems that the trend lines of the curves of ship parameters is almost linear. As assumption, the equations of the trend line are set to be linear. The equations of linear trend line of the resistance and stability parameters are presented at Table 3. The parameters seakeeping (periods of rolling, heaving and pitching) are not evaluated further here. Especially for rolling period, the values are smaller (≈ 3 seconds) than what is expected which is about 12 seconds for the passenger ferries. This phenomena exists mostly for passenger ships where special treatment would be introduced later in this study.

Parameters	Equations of linear trend line
Resistance (*V*s: 19,98 kt)	$f(x) = -2.6267x + 177.45$
Stability criterion:	**Departure**
Angle steady hell ≤ 16	$f(x) = -0.1712x + 8.972$
Angle s.h/m.l im. ≤ 80	$f(x) = -0.7959x + 39.94$
Area1/Area2 ≥ 100	$f(x) = 16.631x - 368.65$
Area 0-GZ_{max} ≥ 0.055	$f(x) = 0.0158x - 0.291$
Area 30 to 40 ≥ 0.03	$f(x) = 0.0059x - 0.0872$
Angle GZ_{max} ≥ 15	$f(x) = 1.8814x - 28.881$
Initial GMt ≥ 0.15	$f(x) = 0.1053x + 0.0686$
Angle pax crowd ≤ 10	$f(x) = -0.3326x + 17.19$
Angle hs turning ≤ 10	$f(x) = -0.2418x + 10.51$
Angle wind heel ≤ 16	$f(x) = -0.1614x + 8.655$

Table 3a: The equations of linear trend line for the resistance and stability parameters (departure condition)

Parameters	Equations of linear trend line
Resistance (*V*s: 19,98 kt)	$f(x) = -2.6267x + 177.45$
Stability criterion:	**Arrival**
Angle steady hell ≤ 16	$f(x) = -0.2264x + 11.08$
Angle s.h/m.l im. ≤ 80	$f(x) = -0.9613x + 46.573$
Area1/Area2 ≥ 100	$f(x) = 16.496x - 407.64$
Area 0-GZ_{max} ≥ 0.055	$f(x) = 0.075x - 0.3876$
Area 30 to 40 ≥ 0.03	$f(x) = 0.0062x - 0.1126$
Angle GZ_{max} ≥ 15	$f(x) = 1.4948x - 21.735$
Initial GMt ≥ 0.15	$f(x) = 0.1029x - 0.0795$
Angle pax crowd ≤ 10	$f(x) = -0.5099x + 23.952$
Angle hs turning ≤ 10	$f(x) = -0.2828x + 12.186$
Angle wind heel ≤ 16	$f(x) = -0.2166x + 10.745$

Table 3b: The equations of linear trend line for the resistance and stability parameters (arrival condition)

The objective of the study is to find the minimum resistance of the modified ship. Here the length of ship (L_{OA}) becomes a single control variable (x). Set the length as control variable x and the resistance as objective function f(x) then the solution of the optimization problem is stated as:

$$\text{Minimize: } f(x) = -2.6267x + 177.45 \qquad (11)$$

In fact, another important objective beside minimizing the resistance is minimizing the ship length. In this study some hull parameters are kept constant (beam, midship area) and other parameters are changed (length, draft, displacement, block and prismatic coefficients, half angle of entrance) then the resistance decreases as the length increases as seen at the resistance curve in Figure 3. Then, to find the minimum resistance based on minimum length, the solution of the optimization problem is stated as:

$$\text{Maximize: } f(x) = -2.6267x + 177.45 \quad (12)$$

Subject to the constraints:

(for departure condition)

$-0.1712x + 8.972$	≤ 16
$-0.7959x + 39.94$	≤ 80
$16.631x - 368.65$	≥ 100
$0.0158x - 0.291$	≥ 0.055
$0.0059x - 0.0872$	≥ 0.03
$1.8814x - 28.881$	≥ 15
$0.1053x + 0.0686$	≥ 0.15
$-0.3326x + 17.19$	≤ 10
$-0.2418x + 10.51$	≤ 10
$-0.1614x + 8.655$	≤ 16
x	> 0

(for arrival condition)

$-0.2264x + 11.08$	≤ 16
$-0.9613x + 46.573$	≤ 80
$16.496x - 407.64$	≥ 100
$0.075x - 0.3876$	≥ 0.055
$0.0062x - 0.1126$	≥ 0.03
$1.4948x - 21.735$	≥ 15
$0.1029x - 0.0795$	≥ 0.15
$-0.5099x + 23.952$	≤ 10
$-0.2828x + 12.186$	≤ 10
$-0.2166x + 10.745$	≤ 16
x	> 0

Using excel solver to solve the problems, it was found that for departure condition, the minimum length of ship (L_{OA}) is 28.18 m. Meanwhile the minimum length of the ship for arrival condition is 30.77 m. In fact, for the arrangement of the ship, the minimum length may fit the nearest length based on the arrangement. Therefore, based on the stability of the ship at arrival condition, the length of the ship was selected for L_{OA} = 31.03 m to fit the percentage of 66% passenger at the main deck.

6. CONCLUSION & RECOMMENDATION

6.1. CONCLUSION

- Since there is lack of available database provided for optimization of medium-speed passenger ferry, then the results of parent ship design provides some relevant ship data during the modification process.
- As the ship length increases then, the draft, hull coefficients, half angle of entrance and ratio A_T/A_M decrease. The decreasing of those parameters contributed to decrease of ship resistance though the wetted surface area increases due to the length
- In general, the longer the ship the less resistance or power meanwhile the stability parameters are increasing and seakeeping parameters are decreasing. The narrower the ship the less resistance or power meanwhile the stability parameters are decreasing, the rolling period is increasing, but pitching and heaving periods are decreasing. The decreasing of ship beam contributes so much for the stability and seakeeping
- For the input design of 254 passengers with the constant beam of 7.00 m and other constant parameters (deck height, midship and transom sections) then the minimum length of the ship is 30.70 m and a little increase to 31.03 m to fit the configuration of 66% passenger at the main deck.
- The longer the ship the less resistance or power of the ship, however the selection of lower length also benefits for construction and operational costs. Then the minimum resistance and other parameters were found due to the selected length.
- In fact, the optimization process this study is conducted for only one configuration of ship which is P254S10. Other variations of ship configurations (seat configurations) and draft variation will end-up with other results instead of the results achieved by optimization of P254S10.

6.2. RECOMMENDATIONS

The future works would be recommended for:

- Variation of ship beams and drafts for the modification and optimization process
- Optimizing the hull forms and local refinement to reduce the resistance
- Optimizing the ship structure in order to reduce the structural weight of the ship
- Executing the model tests in order to achieve better results of ship resistance.
- Selecting the proper screw propeller to reduce the engine power

7. ACKNOWLEDGEMENTS

Special thanks to The Directorate General of Higher Education, Department of National Education, Republic of Indonesia for the scholarship contributed in this study.

8. REFERENCES

1. Molland, A. F., Turnock, S. R., and Hudson, D. A., 'Ship Resistance and Propulsion – Practical Estimation of Ship Propulsive Power', *Cambridge University Press, 32 Avenue of the Americas, New York, USA*, 2011.
2. Nicolaysen, K. H., 'CATRIV – WP 4.2', *Working Group ZE VWS (TU Berlin)*, 1999.
3. Larsson, L. and Hoyte C. R. 'Ship Resistance and Flow', The Principles of Naval Architecture Series', J. Randolph Paulling, Editor, *SNAME, 601 Pavonia Avenue, Jersey City, New Jersey 07306,* 2010.
4. Bureau Veritas, 'The Rules for the Classification of High Speed Craft, *Bureau Veritas,* February 2002.

5. Colton, T., 'The Marine Industry', *Chapter 3 - Ship Design and Construction, written by an International Group of Authorities, Thomas Lamb, Editor, SNAME Publication, Jersey City, New York, USA, Vol. 2,* 2003.
6. Watson, D. G. M., 'Practical Ship Design', *Elsevier Ocean Engineering Book Series, Volume I, ELSEVIER,* 1998.
7. Parsons, M. G., 'Parametric Design', *Chapter 11 - Ship Design and Construction, Written by an International Group of Authorities, Thomas Lamb, Editor, SNAME Publication, Jersey City, New York, USA, Vol. 2,* 2003.
8. Schneekluth, H. and Bertram, V, 'Ship Design for Efficiency and Economy', *Butterwoth-Heinemann, Second Edition, Oxford,* 1998.
9. Lewis, E.V., 'Resistance, Propulsion and Vibration', *Principles of Naval Architecture, The Society of Naval Architects and Marine Engineers, 601 Pavonia Avenue, Jersey City, NJ, Vol. 2,* 1988.
10. Mercier, J.A. and Savitsky, D, 'Resistance of Transom-stern Craft in the Pre-planing Regime', *Stevens Institute of Technology, Report 73-1667,* 1973
11. Gale, P. A., 'The Ship Design Process', *Chapter 5 - Ship Design and Construction, Written by an International Group of Authorities, Thomas Lamb, Editor, SNAME Publication, Jersey City, New York, USA, Vol. 2,* 2003.
12. Knox, J., 'Ferries', *Chapter 38 – Ship Design and Construction, Written by an International Group of Authorities, Thomas Lamb, Editor, SNAME Publication, Jersey City, New York, USA, Vol. 2,* 2003.
13. Levander, K., 'Passenger Ships', *Ship Design and Construction Vol. 2, Chapter 37, Writen by an International Group of Authorities, Thomas Lamb, Editor, SNAME Publication, Jersey City, New York, USA, Vol. 2,* 2003.
14. Calhoun, S. R. and Steven, S.C., 'Human Factor in Ship Design', *Ship Design and Construction, Chapter 15, Writen by an International Group of Authorities, Thomas Lamb, Editor, SNAME Publication, Jersey City, New York, USA, Vol. 2,* 2003.
15. Olson, H. A., 'Trends in Modern Ferry Boat Design', *Paper Presented at the February 8 1990 Meeting of The Northern California Section of The SNAME,* 1990.

9. AUTHORS' BIOGRAPHY

Wolter R Hetharia is PhD student University of Liege. Previously he holds the position of lecturer at Faculty of Engineering, Department of Naval Architecture, Pattimura University. He is responsible for the courses of ship design. Several papers have been published concerning this subject.

Andre Hage holds the current position as Professor – University of Liege. He is responsible for Director of Towing Tank, University of Liege. His previous experience includes ship design and hydrodynamics. Many papers have been published concerning those subjects.

Philippe Rigo holds the current position as Professor – University of Liege. He is responsible for Honorary Research Director – FNRS, ANAST, University of Liege. His previous experience includes ship structures and optimization. Many papers have been published concerning those subjects.

A STUDY ON MEDIUM-SPEED MONOHULL PASSENGER FERRIES WITH DIFFERENT HULL MATERIALS

W R Hetharia, Université de Liége, Belgium, **J Nanlohy, E J de Lima and F Gaspersz,** Faculty of Engineering Pattimura University, Ambon, Indonesia

SUMMARY

Recently, the medium-speed monohull passenger ferries have shown their contributions in some maritime countries. The speed range of those ships is in the transition region between conventional ferries and HSC crafts. Since the ship speeds are beyond the hump speed then those ships need a great engine power. The study was conducted in order to collect, analyze and presented the data in this paper. A parent ship of 254 passengers with the hull material of Aluminium was designed and presented in this study. The ship parameters, particularly the engine power, of those ships with different hull materials were compared. Some comments and recommendations were presented for the future works on ship optimization.

NOMENCLATURE

$A_1 .. A_{27}$	Coefficients in equation (1)
A_T	Transom area (m^2)
A_X	Maximum section area (m^2)
B	Molded beam of the submerged hull (m)
C_B	Block coefficient
C_M	Midship coefficient
C_P	Prismatic coefficient
C_{WP}	Waterplane coefficient
C_∇	Beam-loading coefficient
D	Molded depth (m)
Fn	Froude number (nondimensional)
GM_T	Transverse metasenter height (m)
L	Molded ship length (m)
L_{CB}	Longitudinal center of buoyancy (m. %L)
P_B	Brake power (kW)
P_E	Effective power (kW)
R_T	Total resistance (kN)
T	Design molded draft (m)
T_h	Heave natural period (s)
$T\phi$	Roll natural period (s)
T_Θ	Pitch natural period (s)
U	Coefficient in equation (1) = $\sqrt{2i_e}$
V	Ship speed (m/s, knot)
W	Ratio A_T/A_X in equation (1)
X	Ratio $\nabla^{1/3}/L$ in equation (1)
Z	Ratio ∇/B^3 in equation (1)
w	Taylor wake fraction
t	Thrust deduction factor
α_E, i_e	Angle of entrance of load waterline (degree)
η_h	Hull efficiency
η_o	Propeller efficiency
η_r	Relative rotative efficiency
η_s	Seal efficiency
∇	Molded volume at the design waterline (m^3)
Δ	Displacement at the design waterline (t)

1. INTRODUCTION

The marine transportation of passengers in some maritime countries needs a special attention due to the techinical, economical, safety and comfortability aspects. Particularly in the archipelago countries with many islands this mode of transport is highly required to connect the islands. In the real conditions the amount of passengers are not so high therefore operating several small ships is preferred instead of a big ship. Since long time this mode of transport have been supported by passenger and cargo ships, conventional ferries until the period of high speed crafts (HSC) nowadays. The recent conditions show that traveling with a short time is better or it would be said the faster the better. In fact, it is not always the case. Sail faster in the calm seas is better. In the archipelago countries, the seas are not always calm. With this condition, the HSC will face an excessive slamming effects which makes an uncomfort condition for the passengers. When the waves become biger then the HSC should reduce the speeds and plays against the waves. In addition, the HSC still leaves an issue of great engine power and fuel consumption. For these cases, then the operation of medium-speed passenger ferries is prefereably. Those ships operate with the top speeds less than 25 knots.

Even though those ships are called medium-speed, in fact they still need a biger engine power. It was found from a little database that those ships operate at a range of Froude number Fn from 0.55 to 0.80. In fact, this range is beyond the hump speed (Fn > 0.50). Therefore they need a great engine power to maintain the service speeds. The application of the lighter hull materials such as FRP or Aluminium instead of steel for those ships will give some benefits of increasing the payload or speed or decreasing the engine power. In addition, from the collected database it was found also that for a fixed number of passengers (payload) then those ships have a different dimensions and hull forms. These configurations will affects the ship parameters particularly the engine power. A parent ship of 254 passengers with the hull material of Aluminum was designed in this study. The parent ship was modified due to the arrangement of the passengers on board. The results of design is presented and compared with the existing medium-speed ships

2. THE EXISTANCE OF MEDIUM-SPEED SHIPS PASSENGER FERRIES

2.1. THE TERMINOLOGY OF MEDIUM-SPEED SHIPS

The medium-speed ferries are classified as short-sea ferries where they operate as day ferries and travel in the range of shaltered to open waters. They have the service speeds that do not generally exceed 25 knots [1]. Since the range of Froude number Fn of medium-speed is from 0.55 to 0.80 then these ships are clasified as semi-displacement ships. As shown in Figure 1 that the range of Fn for semi-displacement ships is from 0.50 to 1.0 [2]. The term semi-displacement is known as semi-planing also with the range of Froude number Fn is from 0.50 to 1.0 [3]. In addition, Nicolaysen [4] divided this speed range of $0.5 < Fn < 0.75$ for semi-dispalcement and $0.75 < Fn < 1.0$ for pre-planning. According to the definition, High Speed Craft (HSC) is a craft capable of maximum speed, equal to or exceeding: $3.7\nabla^{0.1667}$ (m/sec) where ∇ = volume of displacement corresponding to the design waterline (m^3) [5]. With the range of displacement and speed of the existing ships, it was found that the speed range is greater then the value of $3.7\nabla^{0.1667}$ (m/sec). Therefore from the structure point of view, those kind of ships are classified as HSC. The term medium-speed is mostly used in the maritime fields recently for those ships with the speed up to 25 knots.

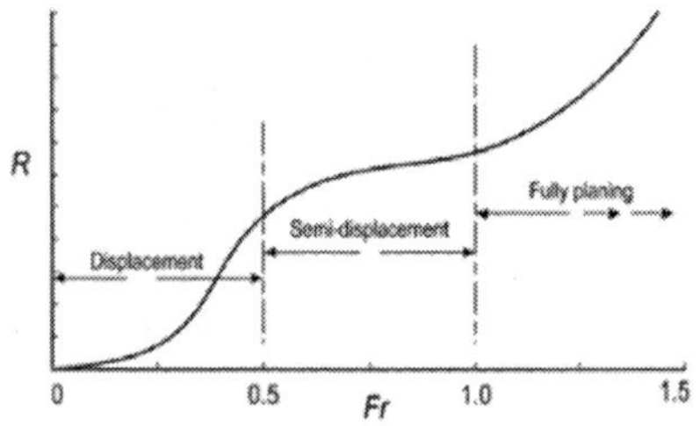

Figure 1: Approximate speed ranges for displacement, semi-displacement and planning craft

2.2. RECENT DEVELOPMENT OF MEDIUM SPEED PASSENGER FERRIES

Medium-speed passenger ferries have been designed and operated in some regions of the world such as Mediteranian, Asia, Pacific, Central America and Europe. Due to their better performance, several shipdesigners, shipbuilders and operators are developing and operating those ferryboats. For example, The Austal Ships [6], The UK ferry operator [7], FBMA Marine [8], Port of Al Khaiman [9], and many more are the ferry operators that take the advantage of those kind of ship. Those ships are operated in short distance voyage where most of them are multi-hulled type (catamarans). However, due to their simplicity, the monohulls are developed and have the future markets. In some regions of Asia, Pacific and Gulf, there are a lot of monohull medium-speed passenger ferries with the hull material of composite (FRP) have been developed and operated. Some pictures of those ships are shown in Figures 2 and 3.

Figure 2: Medium-speed passenger ferries with hull material of Aluminum

Figure 3: Medium-speed passenger ferries with hull material of FRP

2.3. LITERATURE REVIEW OF MEDIUM-SPEED SHIPS

2.3 (a) Resistance and Propulsion of Semi-Displacement Ship.

Resistance data for semi-displacement ships may be found from some sources as shown in References [2, 3, 10, 11]. They may be used as a basis for preliminary power estimates. For semi-displacement ships with hard chine the resistance data may be found from the WUMTIA series [2, 12] and Savitsky pre-planing [10, 11]. In both series of resistance, the of semi-displacement ships depends on the parameters such as length displacement ratio ($L_{WL}/\nabla^{1/3}$), beam-loading coefficient ($C_\nabla = \nabla/B^3$), angle of entrance of the load waterline (i_e) and ratio of transom area to maximum section area (A_T/A_X). The general form of the resistance equation adopted by Mercier and Savitsky is as follows:

$$R_T/W = A_1 + A_2X + A_4U + A_5W + A_6XZ + A_7XU + A_8XW + A_9ZU + A_{10}ZW + A_{15}W^2 + A_{18}XW^2 + A_{19}ZX^2 + A_{24}UW^2 + A_{27}WU^2 \quad (1)$$

where: $X = \nabla^{1/3}/L$; $Z = \nabla/B^3$; $U = \sqrt{2i_E}$;
$W = A_T/A_X$.

The values of the coefficients A_1 to A_{27} and correction factors are presented in the reference [10]. This method is provided in the Maxsurf software. Meanwhile the method derived by WUMTIA (Wolfson Unit for Marine Technology and Industrial Aerodynamics) presented the power prediction as follows:

$$P_E = 453.8 \times \Delta \times (V^2_K / \sqrt{L}) \times (1/C^2_{FAC}) \quad (2)$$

where: Δ = displacement; V_k = ship speed in knots; L = ship length

The constant values of C^2_{FAC} may be found in the Reference [2].

The effective power of the ship is computed as:

$$P_E = R_T \times V \quad (3)$$

where: R_T = total resistance

V = speed of the ship

The engine power (brake power P_B) is computed in relation with the effective power P_E [13] is as follows:

$$P_B = P_E / (\eta_h \eta_o \eta_r \eta_s \eta_b \eta_t) \quad (4)$$

where:

η_h = hull efficiency;

η_o = propeller efficiency;

η_r = relative rotative efficiency = 1.0

η_s = seal efficiency;

η_b = line shaft bearing efficiency;

η_t = transmission efficiency;

$\eta_s \eta_b$ = 0.97 for machinery amidships

η_t = 0.975 for medium speed diesel plant

Hull efficiency is computed as:

$$\eta_h = (1 - t)/(1 - w) \quad (5)$$

where:

w = Taylor wake fraction $= 0.5\, C_B - 0.05$ (6)

C_B = block coefficient

t = thrust deduction factor $= 0.6\, w$ (7)

The maximum continuous rating (MCR) of the main engine is determined by adding a power service margin as 10% to the brake power.

$$\text{MCR} \geq (1 + MS)\, P_B \quad (8)$$

where: MS = power service margin

Two screw propeller units are used for the parent ship. The screw propellers are evaluated based on the propeller data from the Wageningen B-Screw Series [10]. The propeller types of B 4-55, B 4-70, B 4-85 and B 4-100 are evaluated for a range of ship speed from 19 to 21 knots. In addition, an evaluation for the cavitations of the propeller is executed based on Burril Diagram of cavitations. The trend line of suggested upper limit for merchant ship propeller is used in this evaluation. In fact, this trend line is still subjected to back cavitations of propeller blade (less than 5%).

2.3 (b) Stability Requirements for Medium-Speed Passenger Ships

The stability criteria that should be applied for the ship stability are based on HSC Code 2000 MSC 97(73)-Annex 8 Monohull Intact, HSC Code 2000 Chapter 2 Part B Passenger Craft Intact and IMO MSC 36(63) HSC Code Monohull. Those criteria include:

Angle steady hell ≤ 16 deg

Angle steady hell/margin line immersion ≤80 %

Area1/Area2 ≥100 %

Area 0-GZ_{max} ≥0.055 m.rad

Area 30 to 40 ≥ 0.03 m.rad

Angle GZ_{max} ≥ 15 deg.

Initial GMt ≥ 0.15 m

Angle pax crowd ≤10 deg.

Angle high speed turning ≤10 deg.

2.3 (c) Seakeeping Requirements for Medium-Speed Passenger Ships.

The sea keeping parameters were evaluated for rolling, pitching and heaving natural periods. Those are computed due to the following formulas [13].

Roll natural period:

$$T\phi = 2.007\, k_{11} / \sqrt{GM_T} \quad (9)$$

where: $k_{11} = 0.40\, B$

Pitch natural period:

$$T_\Theta = 1.776\, C_{WP}^{-1} \sqrt{(T\, C_B\, (0.6 + 0.36\, B/T)} \quad 10)$$

Heave natural period:

$$T_h = 2.007 \sqrt{(T\, C_B\, (B/3T + 1.2)/C_{WP})} \quad 11)$$

2.4. THE EXISTING DATABASE OF MEDIUM-SPEED PASSENGER FERRIES

Some data and information were collected by the authors concerning the existing monohull medium-speed passenger ferries. It was found that:

- Most of the ship operate at the speed range of 19 to 23 knots.
- Short voyage distances 25 to 200 nautical milles.
- The passenger capacity varies from 100 to 350 passengers.
- The passenger accomodation is provided with the seat.
- The ships generally have two decks i.e. main and upper decks. The passengers are distributed (with certain amount) on those two decks.
- For the ships with the hull material of Aluminium, the ranges of Froude number Fn are 0.55 to 0.80 while for those with the hull material of FRP, the ranges of Fn are 0.53 to 0.62.
- Most of the ships have the hull form of hard chine for both kinds of hull materials of Aluminium and FRP.
- From the collecting data of the existing ships, for a certain number of passengers capacity the ships have the differences in dimensions, speeds and engine powers
- The database of the ships are presented in Figure 9.

3. DESIGN OF PARENT SHIP

It is quite hard to find the detail database from the existing medium-speeds, particularly detail of weight and centers and other data. Therefore a parent ship was designed for the study and future application. Once the parent ship is provided then it can be used to explore the effects of parametric variations [14]. Some important key factors were summarized from the References [14, 15, 16, 17, 18] concerning the arrangement of these passenger ferries. Such factors include the spaces, services, acces, accommodations, facilities on board to ensure the comfort for the passengers and to support the operation of the ship. The arrangement of ship is fixed to fulfill the safety standard regulations. In addition, The design parameters that should be considered during the operation are safety, stability, sea keeping and maneuvering capability. The input design parameters were computed and analyzed during the design process. The process is finished when the outputs meet the ship requirements. The structure components of the ship were determined based on the Rules for the Classification of High Speed Craft [5]. Type of ship is passenger ferry/class B. Type of passenger accommodation: seat in passenger saloon. The hull material of the ship is Aluminum Alloy. The types of alloys used for the ship are 5083 H111 for plating and 6082 T6 for profile. The parameters of parent ship are presented in Table 1. The layout of the ship is shown in Figure 4.

Parameters	Value	Unit
Total number of passengers	254	
Number of seats in row	10	
Number of crews	5	
Number of pax main-upper decks	177-77	
Pax distributions main-upper decks	70-30	%
Service speed	19.98	knots
Navigation range	200	n.m
Length overall, L_{OA}	32.00	m
Length of waterline, L_{WL} (= L_{BP})	29.00	m
Ship beam (at main deck), B	7.00	m
Ship beam (waterline), B_{WL}	6.69	m
Draft, T	1.388	m
Deck height, D	2.600	m
Displacement, Δ	105.3	Tone
Volume displacement ∇	102.7	m^3
Block coefficient, C_B	0.382	
Midship coefficient, C_M	0.547	
Prismatic coefficient, C_P	0.697	
Water plane coefficient, C_{WP}	0.76	
Long. C. of buoyancy, Lcb (fr AP)	48.33	% Lwl
Vertical center of buoyancy, KB	0.967	m
Radius metacenter, BM_T	5.374	m
Vertical center of gravity, KG	3.039	m
Initial metacenter, GM_T	3.302	m
Total resistance (at 19.98 knots)	93.6	kN
Total engine power (MCR)	2402	hp
Total propulsion efficiency η_t	0.588	
Period of rolling, $T\phi$	2.97	Second
Period of pitching, T_Θ	2.31	Second
Period of heaving, Th	2.65	Second

Table 1: Parameters of the parent ship

Since there database of weight and centers for this kind of ship are quite difficult to find then the total weight of the ship is computed and estimated directly from the parent ship. The weights and their centers are presented in Table 2. A margin weight of 4% is added to the ship lightweight. In addition, a margin of VCG of 0.150 m was added for the VCG of the ship lightweight [13].

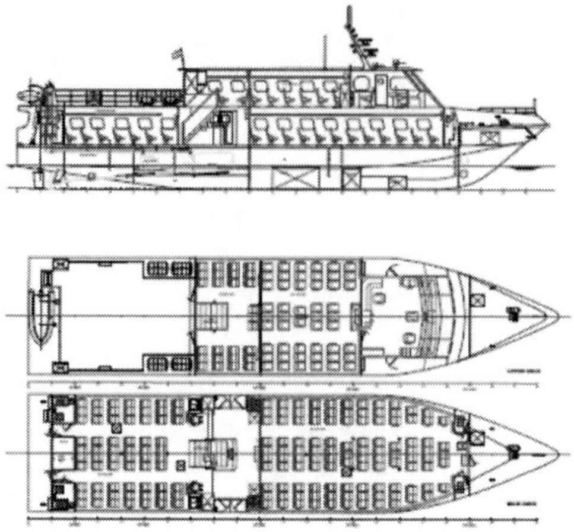

Figure 4: Layout of the parent ship

Item weight	Weight (tone) (tone)	LCG (m)	VCG (m)
Structural weight	38.902	-0.628	2.882
Mach. & system	15.725	-2.900	1.418
Ship outfits	13.960	-0.381	3.773
Sum llightweight	68.587	-1.099	2.728
Margin weight	2.743	-1.099	2.728
Deadweight	33.922	0.764	3.355
Total weight	105.252	-0.498	2.930

Table 2: Weight and centers of the parent ship

Three methods for computation of resistance are presented here, i.e. Holtrop, WUMTIA and Savitsky pre-paling. The result of the resistance is shown in Figure 5. From the figure, it was found that the method of Holtrop was out of range and the other two methods are in good pattern. For the next computation, the resistance is computed based on Savitsky pre-planning method.

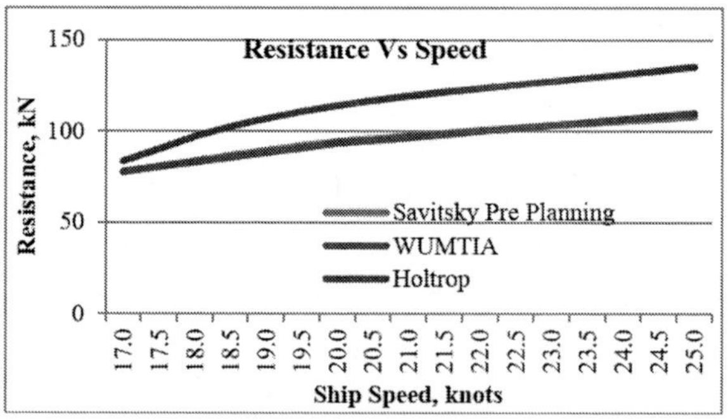

Figure 5: The resistance of parent ship for three methods.

Two units for the main engines are selected for the propulsion system of the ship. It would be better to select the existing types of main engine to be used for the ship. However, as assumption for this preliminary study, the main engine and following characteristics were selected.

The screw propeller with four types are selected and evaluated in this study which are B4-55, B4-70, B4-85, B4-100.

- Type of main engine: MTU Marine Diesel Engine 10V 2000 M70
- Rated power : 1205 bhp
- Rated speed : 2250 rpm
- Gearbox : ZF 3000, i = 2.0
- Type of screw propeller : B 4-70 (BAR: 0.70, Number of blade: 4, P/D: 0.799, Diameter:1.078 m Efficiency: 0.584)
- Speed maximum for the ship : 19.98 knots

4. MODIFICATION AND OPTIMIZATION OF THE PARENT SHIP

4.1. MODIFICATION OF THE PARENT SHIP

The results of ship design are ship dimensions, geometrical hull forms and other design parameters. In fact, modification of hull dimensions and forms would change the design parameters such as powering, cost, stability and others [13, 19, 20]. In fact the modification of hull dimensions and forms may be applied further for the optimization process due to the required optimization objectives. The resistance of semi-displacement shis depends on the parameters of length displacement ratio $L_{WL}/\nabla^{1/3}$ and beam-loading coefficient $C_\nabla = \nabla/B^3$. In fact in the real conditions those parameters are depend mainly on the ship layout. Since the accommodation for the passengers is provided with the seats, then the seat arrangement is evaluated here. The parent ship has 254 passengers. This parent ship has a row of 10 seats due to the ship beam. The ship was named for P254S10. The passenger distibution on main deck is 70% and upper deck is 30 %. Futhermore the modification was executed by shifting one seat row from main deck to upper deck or vice versa. This gives the distribution of passenger main deck of 54, 58, 62, 66, 70, 74, 78, 82 %. With this condition, it was found that the beam of ship is constant (7.0 m) while the length and other design parameters are change.

It is noticed that the changing was done for the shipment of passengers. In this the weight (DWT) is constant and the changing is just for the centers of gravity. In addition, there is a little changing of structural weight. As a result the total ship weight or displacement of the ship was changed. The parameters of ship resistance and stability were computed by using Maxsurf software. The criteria that have been used for the ship stability are based on HSC Code 2000 MSC 97(73) - Annex 8 Monohull Intact, HSC Code 2000 Chapter 2 Part B Passenger Craft Intact and IMO MSC 36(63) HSC Code Monohull Intact. In addition, the ship resistance was computed based on the statistical method of Savitsky pre-planning. The results of the modifications are shown in Figures 6 to 8.

The figures 6 and 7 show that as the length increase then the average total resistance or engine power decrease 2.7 %. From the Figure 8 it is seen that the initial metacenter increases as length increases. The average value of increasing initial metacenter is 3.6 %.

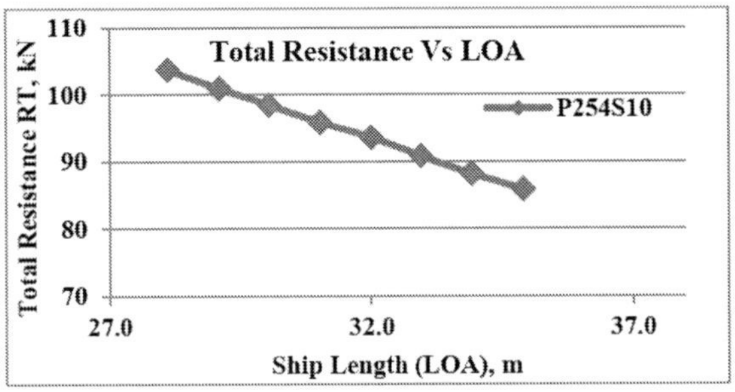

Figure 6: Total resistance due to ship length

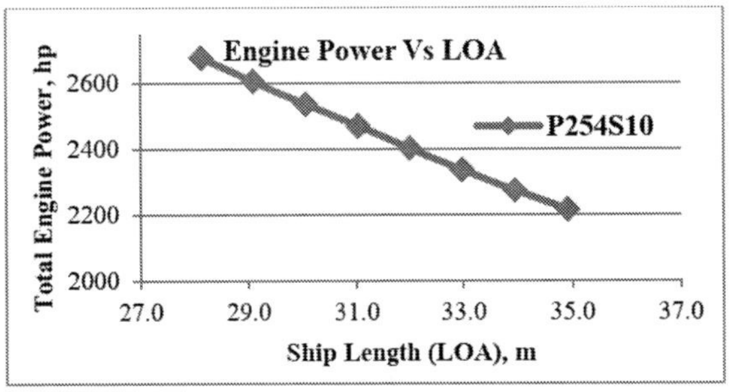

Figure 7: Total engine power due to ship length

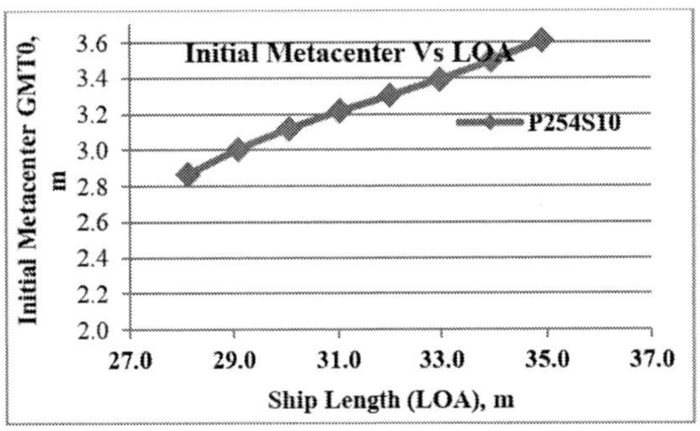

Figure 8: Initial metacenter due to ship length

4.2 OPTIMIZATION PROCCESS FOR PARENT SHIP

Since the modification of the ship was done due to the ship length, then all parameters of resistance, stability and seakeeping are presented as the function of ship length. From Figures 6 to 8 it seems that the trend lines of the curves of ship parameters is almost linear. As assumption, the equations of the trend line are set to be linear. The equations of linear trend line of the resistance and stability parameters are presented at Table 3 (for arrival condition). The parameters seakeeping (periods of rolling, heaving and pitching) are not evaluated further here. Especially for rolling period, the values are smaller

(≈ 3 seconds) than what is expected which is about 12 seconds for the passenger ferries.

Parameters	Equations of linear trend line
Resistance (Vs: 19,98 kt)	$f(x) = -2.6267x + 177.45$
Stability criterion:	**Arrival**
Angle steady hell ≤ 16	$f(x) = -0.2264x + 11.08$
Angle s.h/m.l im. ≤ 80	$f(x) = -0.9613x + 46.573$
Area1/Area2 ≥ 100	$f(x) = 16.496x - 407.64$
Area 0-GZ_{max} ≥ 0.055	$f(x) = 0.075x - 0.3876$
Area 30 to 40 ≥ 0.03	$f(x) = 0.0062x - 0.1126$
Angle GZ_{max} ≥ 15	$f(x) = 1.4948x - 21.735$
Initial GMt ≥ 0.15	$f(x) = 0.1029x - 0.0795$
Angle pax crowd ≤ 10	$f(x) = -0.5099x + 23.952$
Angle hs turning ≤ 10	$f(x) = -0.2828x + 12.186$
Angle wind heel ≤ 16	$f(x) = -0.2166\,x + 10.745$

Table 3: The equations of linear trend line for the resistance and stability parameters (arrival condition)

The objective of the study is to find the minimum resistance of the modified ship. Here the length of ship (L_{OA}) becomes a single control variable (x). Set the length as control variable x and the resistance as objective function f(x) then the solution of the optimization problem is stated as:

$$\text{Minimize: } f(x) = -2.6267x + 177.45 \qquad (11)$$

In fact, another important objective beside minimizing the resistance is minimizing the ship length. In this study some hull parameters are kept constant (beam, midship area) and other parameters are changed (length, draft, displacement, block and prismatic coefficients, half angle of entrance) then the resistance decreases as the length increases as seen at the resistance curve in Figure 3. Then, to find the minimum resistance based on minimum length, the solution of the optimization problem is stated as:

$$\text{Maximize: } f(x) = -2.6267x + 177.45 \qquad (12)$$

Subject to the constraints:

(for the arrival condition)

$$-0.2264x + 11.08 \leq 16$$
$$-0.9613x + 46.573 \leq 80$$
$$16.496x - 407.64 \geq 100$$
$$0.075x - 0.3876 \geq 0.055$$
$$0.0062x - 0.1126 \geq 0.03$$
$$1.4948x - 21.735 \geq 15$$
$$0.1029x - 0.0795 \geq 0.15$$
$$-0.5099x + 23.952 \leq 10$$
$$-0.2828x + 12.186 \leq 10$$
$$-0.2166\,x + 10.745 \leq 16$$
$$x > 0$$

Using excel solver to solve the problems, it was found that for departure condition, the minimum length of ship (L_{OA}) is 28.18 m. Meanwhile the minimum length of the ship for arrival condition is 30.77 m. In fact, for the arrangement of the ship, the minimum length may fit the nearest length based on the arrangement. Therefore, based on the stability of the ship at arrival condition, the length of the ship was selected for L_{OA} = 31.03 m to fit the percentage of 66% passenger at the main deck.

5. COMPARISSON OF THE EXISTING SHIPS

After executing the optimization process for the parent ship, the rest of ship configurations presented in Table 4 are satisfy the criteria of stability.

Parameters	Values				
Pax on main deck (%)	66	70	74	78	82
L_{OA} (m)	31.03	32.00	32.97	33.94	34.91
R_T (kN)	95.87	93.95	90.78	88.19	85.85
Tot Eng Power (hp)	2470	2402	2337	2273	2215
GMt (m)	3.22	3.302	3.391	3.50	3.61
Angle $GZmax$	29.50	31.00	32.90	35.00	37.10
Area 0 – 30 (m.rad)	0.095	0.099	0.105	0.110	0.119
Tr (second)	3.01	2.97	2.92	2.87	2.82

Table 4: Parameters of the parent ship satisfied the stability criteria

Ship Name	Material	Tot. Pax	Speed	Tot Power	LOA	B	T
			knot	hp	m	m	m
278 pax	FRP	278	20	3000	35.85	6.8	1.25
250 pax	FRP	250	21	3000	34.58	6.7	1.09
245 pax base ship	Alu.	245	19.7	2142	30.00	7.0	1.95
243 pax	Alu.	243	24.0	2490	30.00	6.4	1.80
267 pax	Alu	267	23.0	2514	32.00	7.4	1.25
Parent A	Alu.	254	19.98	2470	31.03	7.0	1.39
Parent B	Alu.	254	19.98	2215	34.91	7.0	1.38
Comparisson to the base ship							
278 pax	FRP	11.87	1.5	28.6	16.32	-2.94	-56
250 pax	FRP	2	6.19	28.6	13.25	-4.48	-78.9
243 pax	Alu	-0.82	17.92	13.98	0	-9.37	-8.33
267 pax	Alu	8.24	14.35	14.80	6.25	5.41	-56
Parent A	Alu	3.54	1.4	13.3	3.32	0	-40.4
Parent B	Alu	3.54	1.4	3.3	14.07	0	-42.6

Table: 5 comparisson of the power and speed of the existing ships.

It is quite hard to find the the ship with the same capacity of passengers, speed and dimensions. However it may be seen from the Figure 9 and Table 5 that for the ships with the number of passengers and speed which are almost similar they have the different dimensions. This may affect the total power of the ship. In Table 5, some ships with the number of passengers and speeds which are almost similar are presented and compared. Even the parameters are almost the same in fact their dimensions and power are quite different. A base ship of 245 pax with speed 19.7 knot was choosen. Then, other ships are compared to the base ship. It may be seen that a small change of number of passengers and speed has a great effect on the engine power.

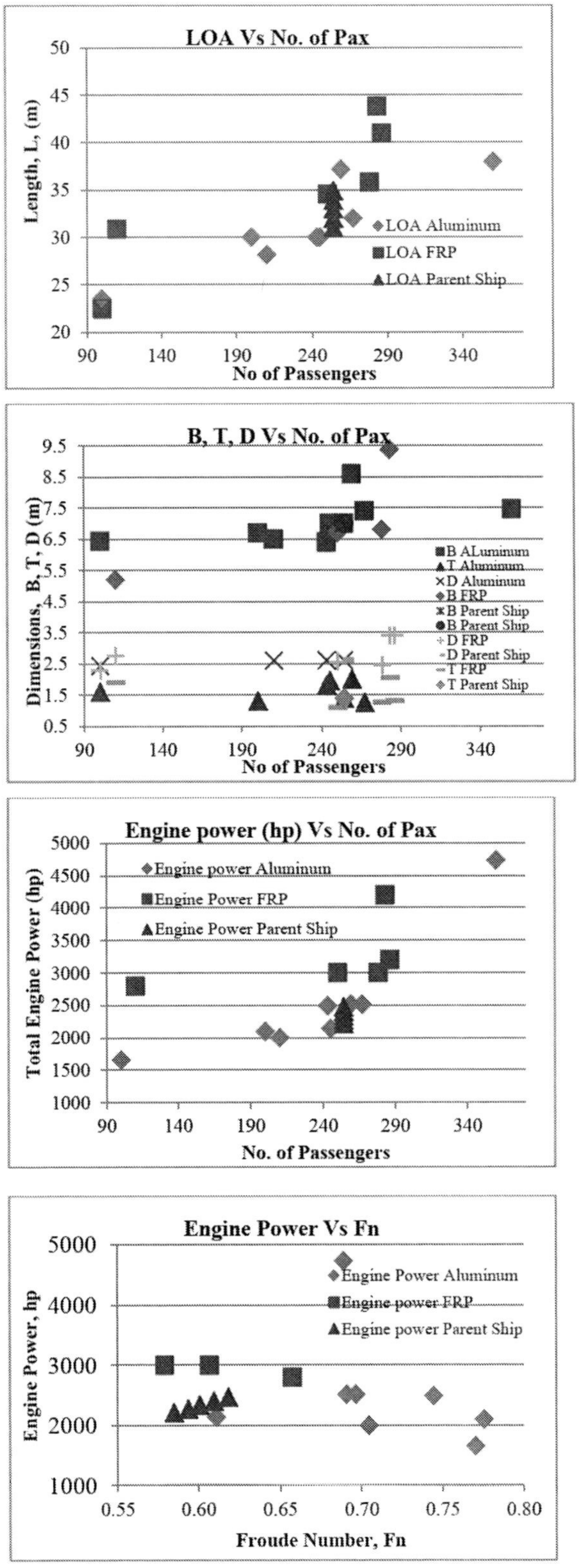

Figure 9: Comparisson of the existing medium-speed passenger ferries

From the existing ships it is seen in Figure 9 also that generaly the ships with hull material of FRP have a greater power even their speeds are smaller compared to the ship with the hull material of aluminium. In addition, their dimensions are slightly greater than those of the ships with hull material of Aluminum.

Meanwhile, the modifications of design ships are in good pattern in the range of Aluminum ships. In this case the engine power would be minimize up to certain value but the length is greater. In fact, the future study with variation of ship dimensions and hull forms would give a better results instead of the existing optimization with one configuration.

6. CONCLUSION & RECOMMENDATION

6.1 CONCLUSIONS

- It may be seen that for a given number of passengers the dimensions of the ships is different in length, beam and draft. Therefore this affect the power required for the ships.
- Generally the ship with hull material of FRP have the great engine power.
- The existance of parent ship is in the range of Aluminum ships, however, with the futher modifiaction of parent ship dimensions then it may give a better result.
- A proper selecting of ship arrangement (lay out) would give a better performance of the ship.
- The arrangement of parent ship and its modification are in good pattern in the range of Aluminium ships however the future study of optimization would give a better results instead of the existing optimization of parent ship.

6.2. RECOMMENDATIONS

The future works would be recommended for:

- An intensive study on the layout and geometrical hull forms of the existing ships to give a better performance of the ship.
- Study concerning other ship parameters particularly the stability of the ships at large angle of inclination.

7. ACKNOWLEDGEMENTS

Special thanks to The Directorate General of Higher Education, Department of National Education, Republic of Indonesia for the scholarship contributed in this study.

8. REFERENCES

1. Colton, T., 'The Marine Industry', *Chapter 3 - Ship Design and Construction, written by an International Group of Authorities, Thomas Lamb, Editor, SNAME Publication, Jersey City, New York, USA, Vol. 2*, 2003
2. Molland, A. F., Turnock, S. R., and Hudson, D. A., 'Ship Resistance and Propulsion – Practical Estimation of Ship Propulsive Power', *Cambridge University Press, 32 Avenue of the Americas, New York, USA*, 2011.
3. Larsson, L. and Hoyte C. R. 'Ship Resistance and Flow', The Principles of Naval Architecture Series', J. Randolph Paulling, Editor, *SNAME, 601 Pavonia Avenue, Jersey City, New Jersey 07306*, 2010.
4. Nicolaysen, K. H., 'CATRIV – WP 4.2', *Working Group ZE VWS (TU Berlin)*, 1999.
5. Bureau Veritas, 'The Rules for the Classification of High Speed Craft, *Bureau Veritas*, February 2002.
6. Significant Small Ships, RINA Publication, 2003
7. Ship and Boat International, RINA Publication, March/April 2009
8. FBMA Marine Inc, Orkney 3rd, September 2007
9. Work Boat World, January 2006
10. Lewis, E.V., 'Resistance, Propulsion and Vibration', *Principles of Naval Architecture, The Society of Naval Architects and Marine Engineers, 601 Pavonia Avenue, Jersey City, NJ, Vol. 2*, 1988.
11. Mercier, J.A. and Savitsky, D, 'Resistance of Transom-stern Craft in the Pre-planing Regime', *Stevens Institute of Technology, Report 73-1667*, 1973
12. Robinson, J. L., 'Performance prediction of chine and round bilge hull forms', *International Conference: Hydrodynamics of High Speed Craft, Royal Institution of Naval Architects, London*, November 1999
13. Parsons, M. G., 'Parametric Design', *Chapter 11 - Ship Design and Construction, Written by an International Group of Authorities, Thomas Lamb, Editor, SNAME Publication, Jersey City, New York, USA, Vol. 2*, 2003
14. Gale, P. A., 'The Ship Design Process', *Chapter 5 - Ship Design and Construction, Written by an International Group of Authorities, Thomas Lamb, Editor, SNAME Publication, Jersey City, New York, USA, Vol. 2*, 2003.
15. Knox, J., 'Ferries', *Chapter 38 – Ship Design and Construction, Written by an International Group of Authorities, Thomas Lamb, Editor, SNAME Publication, Jersey City, New York, USA, Vol. 2*, 2003.
16. Levander, K., 'Passenger Ships', *Ship Design and Construction Vol. 2, Chapter 37, Written by an International Group of Authorities, Thomas Lamb, Editor, SNAME Publication, Jersey City, New York, USA, Vol. 2*, 2003.
17. Calhoun, S. R. and Steven, S.C., 'Human Factor in Ship Design', *Ship Design and Construction, Chapter 15, Written by an International Group of Authorities, Thomas Lamb, Editor, SNAME Publication, Jersey City, New York, USA, Vol. 2*, 2003.
18. Olson, H. A., 'Trends in Modern Ferry Boat Design', *Paper Presented at the February 8 1990 Meeting of The Northern California Section of The SNAME*, 1990.
19. Watson, D. G. M., 'Practical Ship Design', *Elsevier Ocean Engineering Book Series, Volume I, ELSEVIER*, 1998.
20. Schneekluth, H. and Bertram, V, 'Ship Design for Efficiency and Economy', *Butterwoth-Heinemann, Second Edition, Oxford*, 1998.

9. AUTHORS' BIOGRAPHY

Wolter R Hetharia is PhD student University of Liege. Previously he holds the position of lecturer at Faculty of Engineering, Department of Naval Architecture, Pattimura University. He is responsible for the courses of ship design. Principal author: A Modification of Skipjack Pole and Line as Multi-Purpose Fishing Vessel (ICSOT 2010, Surabaya, Indonesia). Principal author: From Kora-Kora to Modern Trimaran: A Design Prototype of Small Passenger Trimaran for Moluccas Waters (Proceedings of National Conference Departmernt of Marine Affairs, Republic of Indonesia 2010, Ambon). Principal Author: Design and Operation of small unsikable fishing boat for local fishermen (Proceedings of National Conference Departmernt of Marine Affairs, Republic of Indonesia 2010, Ambon).

J Nanlohy is the lecturer at The Faculty of Engineering, Department of Naval Architecture, Pattimura University. He is responsible for the course of ship propulsion. Author: From Kora-Kora to Modern Trimaran: A Design Prototype of Small Passenger Trimaran for Moluccas Waters (Proceedings of National Conference Departmernt of Marine Affairs, Republic of Indonesia 2010, Ambon).

E J de Lima is the lecturer at The Faculty of Engineering, Department of Naval Architecture, Pattimura University. He is responsible for the course of ship design. Author: From Kora-Kora to Modern Trimaran: A Design Prototype of Small Passenger Trimaran for Moluccas Waters (Proceedings of National Conference Departmernt of Marine Affairs, Republic of Indonesia 2010, Ambon).

F Gaspersz is the lecturer at The Faculty of Engineering, Department of Naval Architecture, Pattimura University. He is responsible for the course of fast ship. Author: Design and Operation of small unsikable fishing boat for local fishermen (Proceedings of National Conference Departmernt of Marine Affairs, Republic of Indonesia 2010, Ambon).

*** Formatting Issue - Best Available Paper/Graphic ***

DEVELOPING A LOW COST VEHICLE/PASSENGER FERRY IN RESPONSE TO THE INCREASED COMPETITION FROM AIR TRAVEL IN SOUTH-EAST ASIA

J C Knox and **C M Evans**, Lightning Naval Architecture, Australia

SUMMARY

The transportation of people around South-East Asia, particularly Indonesia, is a changing market. The growth of low cost airlines in the region is bringing the option to fly within the financial means of more and more people taking business away from traditional passenger ferry routes and operators. At the same time vehicle ownership and demand for consumer goods is increasing. Lightning Naval Architecture believe that these trends indicate the need for more Passenger/Vehicle Ferries which provide increased revenue streams from those carrying only passengers. This paper discusses the development of a low cost ferry design for routes of between 150nm and 300nm in Indonesia. Whilst the build and operating cost are of primary concern, the design also looks to address widely reported safety concerns with Ro-Ro vessels operating in the Archipelago

1. INTRODUCTION

Air travel in Indonesia grew by 15% in the first Quarter of 2012 [1], a pattern being repeated around South-East Asia. Whilst tourism may have stimulated this growth, it has widely been reported in the maritime press that cheaper air fares will see more of the local population travelling domestically by air rather than sea. Coupled with concerns identified by IMO and Interferry [2] over safety standards on ferries in the developing world, there is a dual threat to ferry markets in the region.

Lightning Naval Architecture (LNA), a Sydney based Ship Design Consultancy believe that a different approach to ferry operations in the region is required. With the increase in air travel and falling passenger revenues, a natural response is to look for vehicle carrying capacity to increase revenues, particularly when the Indonesian Government predicts a 200% growth in car ownership and 50% increase in road freight by 2025 [3]. In such a highly populated developing region however; passengers will remain a very important source of revenue.

LNA therefore presents a conceptual day ferry design to respond to the changing travel market in Indonesia. At 70.2 LBP with facilities for 670 passengers it retains a high passenger capacity for the size of vessel whilst providing 280 lane metres of heavy vehicles. Targeted at 150-300nm routes it has facilities designed to match the market and anticipated 12-20 hour crossing in either a single or two class configurations. The design utilises a simple hullform and arrangement to reduce construction and operating costs, whilst eliminating design elements which have provided safety concerns in existing vessels.

2. INDONESIAN FERRY MARKET REVIEW

2.1 Routes and Trade

Indonesia is an Archipelago of some 17,500 islands of which around 1000 are permanently inhabited. The most heavily populated island by some margin is Java where the capital, Jakarta, is located. Besides Java significant populations live on the islands of Sumatra, Kalimantan (Borneo), Sulawesi, Bali and Lombok. Smaller populations are present in the eastern portion of the country. The national ferry company, PELNI, operates the most extensive network in Indonesia through its fleet of over 20 passenger ships. Much of PELNI's fleet are relatively modern passenger only vessels built at Germany's Meyer Werft Shipyard, their capacity for Ro-Ro is, at time of writing, limited to one vessel operating a 2-week schedule on routes between Java, Kalimantan and Sulawesi [4]

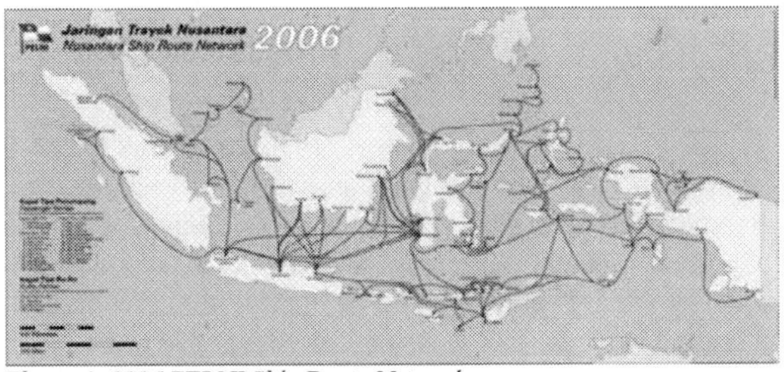

Figure 1: 2006 PELNI Ship Route Network

One of the main operators of Ro-Ro vessels in Indonesia is another state-owned ferry company ASDP [5]. The ASDP network is smaller than that of PELNI and focuses mainly on short-sea crossings between islands, for example Ketapang (Java) to Gilimanuk (Bali), or across bays within islands such as the Bajo'e to Kolaka crossing within Sulawesi. ASDP also operates longer routes between Kalimantan and Sulawesi.

The busiest and most competitive ferry route in Indonesia is the Merak (Java) to Bakauheni (Sumatra) [6] route which is operated by several ferry companies. Competition on this route is intense however with the threat of the Sunda Strait bridge development looming large investment in new tonnage seems unlikely [7].

Information on numbers of people and vehicles carried can be difficult to obtain however it has been suggested that some 60% of vehicles carried on the Merak – Bakauheni route are motorbikes [6].

2.2 Future Ferry Demand

A future Ro-Ro design would source its income from both passenger and vehicle revenue streams. As with ferry operations in the developed world, the transportation of freight is likely to be the key factor to the profitability of any ferry operation. Reliance on freight revenues only seems somewhat risky however in a developing economy and so personal or trade vehicles must also be considered important.

It has already been discussed that car ownership and road freight numbers are predicted to rise substantially by 2025. More detailed statistics on numbers of vehicles are available from the Indonesian statistics office, Badan Pusat Statistik [8]. The website provides a range of high level statistics which can be used to establish general trends.

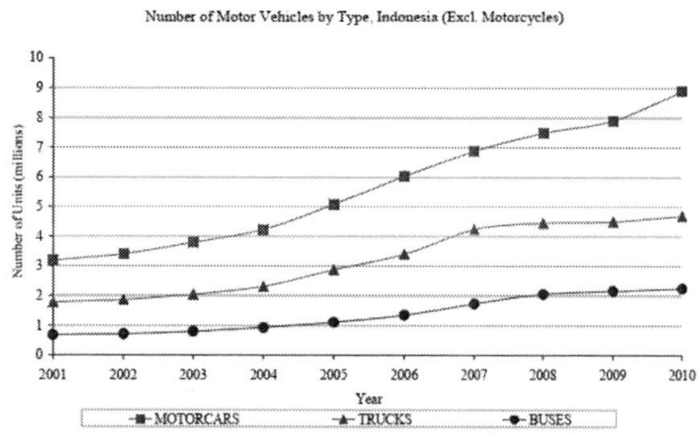

Figure 2: Number of vehicles by type

Figure 2 shows that since 2001 there has been a significant increase in the numbers of Motorcars and Trucks in Indonesia with a smaller increase in Bus numbers. These increases however are considerably smaller than the increase in Motorcycles as shown on Figure 3.

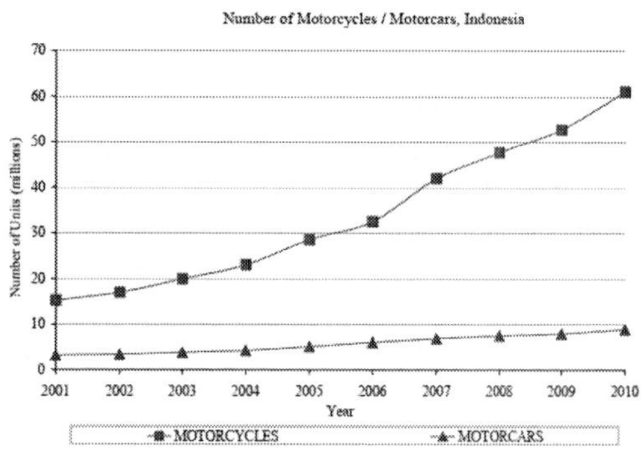

Figure 3: Number of Motorcycles and Motorcars

It would appear therefore that motorcycles will form a significant proportion of the numbers of vehicles carried. This is good news for the demand to transport these vehicles between islands. However since an estimated 75-80% of the population live on either Java or Sumatra, there is a clearly much smaller population and associated share of the vehicular movements on offer to the ferry market when or if an economically viable Sunda Strait Bridge is completed.

If investment in ferries between Java and Sumatra is unattractive whilst the Sunda Strait Bridge is under consideration, there are still considerable populations in Kalimantan (est. 19 million) and the islands of Sulawesi (est. 16 million) and Bali (est. 4 million). These destinations are already served by existing Ro-Ro routes (PELNI and ASDP) either between these islands or from Java to one or more of the islands.

Badan Pusat Statistik has statistics on the numbers of domestic passenger journeys undertaken in Indonesia over the 8 years 2003-2010 by both air and sea.

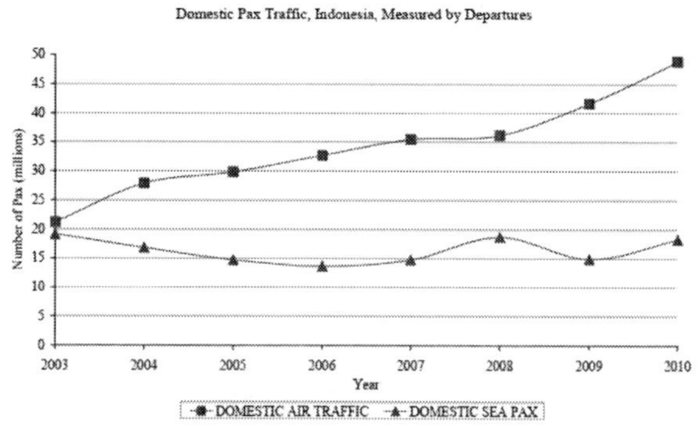

Figure 4: 8 year Passenger Departures by Air and Sea

Figure 4 indicates that whilst passenger numbers travelling domestically by air annually has more than doubled from 21 to 48 million journeys, the numbers of passengers travelling by sea has oscillated between 15 and 20 million journeys. This supports the argument that the ferry industry faces increased competition from airlines which may be stunting the growth in journeys undertaken by sea. Ferries do still have an important role to play in the transportation of passengers; however with increased vehicle numbers the opportunity to carry vehicles as well as passengers is attractive.

2.3 Location of Future Ferry Demand

Determination of where future demand (the route it will be on) for new ferry tonnage is a complicated decision outside of the scope of this paper however some publicly available information gives an indication of where demand may exist. Badan Pusat Statistik has some information on the trade at Indonesia's 5 largest ports for domestic trade which are of interest. The largest ports in Indonesia in terms of handling domestic cargo are as follows:

- Tanjung Priok (Jakarta, Western Java)
- Balikpapan (Eastern Kalimantan)

*** Formatting Issue - Best Available Paper/Graphic ***

forum to try to improve safety standards on ferries in the developing world. A recently published action plan by that forum highlighted the need for governments to establish "appropriate shipping safety policies and standards (e.g. for purchase and operation of second hand and converted ships, new buildings standards to ensure adequate safety margins on stability);"

Indeed it would appear that many recent incidents involving fatalities on Indonesian ferries are as a result of loss of intact stability. An investigation at the University of New South Wales [10] undertaken by a Masters degree student found that in the majority of cases where a Ro-Ro vessel had capsized it was as a result of water encroaching onto the open vehicle deck during bad weather / large sea states and becoming trapped. The resultant free surface effect and possible downflooding through openings on the vehicle deck, ultimately destroying the vessel's stability leading to sudden capsize.

The ease by which water can enter the vehicle deck can, at least partially, be attributed to the open nature of the vehicle deck design on typical vessels operated in Indonesia. Many vessels are ex-Japanese vessels designed to minimize tonnage (hence port dues) on sheltered coastal routes. These vessels take advantage of the fact that vehicle decks with sufficiently large openings in the side may be excluded from the tonnage calculation. Whilst this is an adequate arrangement for short coastal voyages, it is considered by the authors to be unsuitable for operation on open water for long voyages where the weather can change unexpectedly. The casualty rate would appear to support this conclusion.

Subject to safe operation by the crew and ship's management there should be no reason why these vessels cannot be operated safely on short sheltered routes. It is the longer routes, such as those considered in this paper, which represent the most significant risk.

In our design LNA shall offer some technical solutions to the problem of insufficient intact stability.

2.6 Indonesian Ferry Market Review Conclusion

Indonesia is a vast archipelago and despite increased competition from air travel it seems likely that there will be continued and sustained demand for ferry links between the various islands. With added competition it seems prudent to diversify operations to include the carriage of vehicles, especially when the number of vehicles is increasing strongly.

With the prospect of a bridge across the Sunda Strait in the medium term it would appear that investment would be best made in routes between Java, Kalimantan and Sulawesi where the local populations are likely to have greater demand for manufactured goods and processed food produce.

Such a Ro-Ro vessel should be fit for purpose for the intended route. A crossing time of around 18-24 hours is a considerable distance from land and exposes the vessel to increased risk of encountering storm conditions which have been the likely cause of recent ferry capsizes in the country.

3. RO-RO VESSEL FOR INDONESIA

3.1 Design Brief

The LNA design shall fulfil the following design brief. This is based on the perceived demand from the research in section 2 of this paper.

3.1.1 Intended Route

LNA, as a design consultancy, cannot be certain as to the exact demand anticipated on specific routes however LNA has so far concluded that the development of a design suitable for operation between the ports of Tanjung Perak, Balikpapan and Makassar would offer the best potential for the development of a profitable route. If an economical average voyage speed of 15 knots is assumed then the following voyage schedules are possible with a single-ship operation:

Tanjung Perak – Balikpapan

500nm @ 15 Knots 33hr 20 mins (1 crossing every other day, 3-day return with 1 lay day gives 2 return crossings a week)

Tanjung Perak – Makassar

500nm @ 15 knots 33hr 20 mins (1 crossing every other day, 3-day return with 1 lay day gives 2 return crossings a week)

Balikpapan – Makassar

280 nm @ 15 knots 18hr 40 mins (1 crossing per day, 2-day return with 1 lay day gives 3 return crossings a week)

An alternative to this is to operate a loop incorporating these 3 ports. This would require 2 vessels to deliver a regular service in both directions. Having more than one ship allows an operator to continue to operate a reduced service when one vessel is in dry-dock. Reliability of service is a powerful driver of freight growth.

Determining the exact route is beyond the scope of this report since LNA are not privy to commercial information on existing routes; however with the need for economical operation the average speed of 15 knots is chosen.

3.1.2 Construction

The vessel shall be built in Indonesia so that it can offer benefit to the local economy during construction and when in operation. To reduce building costs the arrangement shall be as simple as possible whilst still providing the necessary facilities for cargo and passengers. The hullform shall be designed to reduce the requirement for large numbers of double curvature plates which considerably add to the cost of construction. The machinery arrangement shall be chosen to provide the most economical solution in terms of initial build and operation with the compromise being convenience in manoeuvring which is less important for a long route with long turnaround time in port. The long turn-around time available negates the need for a bow or side ramp for vessel loading, a single stern ramp is considerably cheaper to construct.

3.1.3 Vessel Capacity

The use of Ro-Ro ships on these routes is an emerging market. For an emerging market a small vessel of between 250 and 300 lane metres would be suitable. This should have a full height deck of 4.6m to allow a full freight cargo to be carried. This full height also provides flexibility for the future installation of mezzanine decks for the carriage of additional small vehicles. Passenger numbers should be the maximum practicably achievable for the resulting vessel particulars. LNA will design the passenger lounges in a single class configuration with the flexibility to segregate into further class divisions to suit an operators requirements.

3.1.4 Regulatory Regime

The ship will be designed to meet the requirements of an IACS classification society. For route flexibility and allowing for the length and remoteness of the routes previously discussed, the ship shall comply with the requirements of SOLAS for a Short International voyage. The design will be sufficiently adaptable to relax from the SOLAS requirements if domestic regulations allow.

The problem of water of deck has been described as a major concern with the safety of Ro-Ro vessels in the region. This report will discuss the stability performance of the design with water on deck. This will be of particular importance if the vehicle deck is required to be open to reduce the tonnage and subsequently port dues. It is the opinion of LNA that for a route of this nature the vehicle deck should be weathertight, i.e. openings in the side shell to reduce tonnage should be avoided.

3.2 Design Overview

A General Arrangement of the LNA design to meet the design brief is presented in Figure 8. This figure is displayed at the end of the paper.

This section of the report outlines some of the features of the design

3.2.1 Principal Particulars

Classification:	100A1Passenger/Vehicle Ferry
Reg. Compliance:	SOLAS Short International Voyage
Length OA:	82.0 metres
Length BP:	70.2 metres
Beam, mld.:	17.4 metres
Draught, design:	3.6 metres
Depth, Vehicle Deck	6.0 metres
Depth, Weather Deck:	11.4 metres
Lane Metres:	280 @ 4.6m clear height
Deadweight:	800 tonnes
Passengers:	670
Crew:	30
Service Speed:	15.5 knots
Normal Range:	1900nm at 15.5 Knots
Range with LR Fuel:	3700nm at 15.5 Knots
Access Arrangement:	15m universal stern ramp

3.2.2 Machinery Arrangement

Main Engines:	2 * Wärtsilä 8L26 (2 * 2720 KW)
Gensets:	3 * Mitsubishi S6R2-MPTA (3 * 500 KW)
Propeller Diameter:	2.7m
Propeller Type:	2 * Fixed Pitch
Fin Stabilisers:	Yes
Bow Thruster:	1 * 600 KW
Anti-Heel System:	Optional
Fuel Oil:	154 tonnes
Additional LR Fuel:	145 tonnes
Fresh Water:	146 tonnes
Water Ballast:	202 tonnes

3.2.3 Hullform

The vessel's beam is chosen to provide four vehicle lanes, a central casing and space for the side shell structure. The length of the ship is what is required to provide the necessary lane metres of vehicles.

The hullform has been designed to minimise the amount of double curvature plates. The aft buttocks are straight and the bilge is a double chine which dispenses with the need for any curvature in this section.

The bow section has mainly single curvature panels with a double chine system. The bulb is knuckled to further reduce the complexity and cost of construction. The only double curvature plates required are at the bulb / stem intersection. Further refinement of the hullform to further reduce the construction cost is certainly possible as the design progresses.

3.2.4 Vessel Layout

The General Arrangement is shown in Figure 8 at the end of this paper. The ship has 7 working decks. This section outlines some of the features of the vessel:

Deck 1 (Tanktop):
- Fore and aft engine rooms
- Engine control room in the forward engine room.
- Stabiliser spaces
- Access from Deck 5

Deck 2 (Tween Deck):
- Provisions stores
- Sewage / Auxiliary space
- Laundry
- Crew fitness / potential crew prayer area
-

Deck 3 (Main Deck):
- Vehicle deck
- Bosun's store
- Main access via stern ramp

Deck 4 (Stair Deck):
- Upper Bosun store
- Potential mezzanine vehicle decks

Deck 5 (Passenger Deck):
- Outside seating (64 seats)
- Inside Lounge (265 seats)
- Cafeteria (176 seats)
- Galley
- Toilets
- Mooring decks
- Potential passenger prayer room

Deck 6 (Crew Deck):
- Outside seating (48 seats)
- Inside Lounge (171 seats)
- Crew Mess
- Crew Cabins

Deck 7 (Bridge Deck):
- Crew Recreation
- Meeting room and Office
- Crew cabins
- Bridge

The layout has been arranged to provide as flexible an arrangement as possible to be tailored by the operator. LNA recognises the probable requirement for prayer facilities onboard for passengers and crew. The recreation room on Deck 2 can provide optional space for crew prayer. Passenger prayer facilities would be best constructed on Deck 5 at the loss of some passenger seating. The Deck 2 area is not currently passenger accessible.

The superstructure has minimum curve and shape to reduce construction cost.

3.2.5 Ship-Shore Interface

LNA recognise the potential lack of dedicated Ro-Ro port facilities at some Indonesian ports. With this in mind the ship has been designed with a single aft ramp for vehicle / passengers / provision and waste handling. This is an economical alternative to multiple dedicated stations or doors to provide each service. The design can operate stern-to on quays from 0.8 to 4 metres height above the waterline using the ramp as shown on the General Arrangement drawing. Once the actual facilities for the intended route are known the actual ramp may be reduced in size and hence construction cost.

3.2.6 Lifesaving Arrangements

The ship has a lifesaving provision in accordance with SOLAS for a short international voyage. This requires, for 700 persons onboard, the carriage of two lifeboats with an aggregate capacity of 210. In this case two no. 106-person lifeboats are chosen. The remainder of the lifesaving capacity is provided by davit launched liferafts mounted on deck 6. Embarkation of the liferafts is from deck 5. Davit launched lifeboats are a more economical alternative to MES systems.

3.3 Vessel Stability

3.3.1 Regulatory Requirements

The vessel complies with the intact stability requirements of IMO Resolution A749(18) for a passenger ship.

Damage stability has been assessed to comply with the requirements of SOLAS 2009 consolidated Part B-1 for 700 persons on board.

3.3.2 Additional Stability for Water on Deck

In addition to the statutory stability requirements for a passenger ship, calculations have been undertaken to demonstrate the vessels ability to remain stable with a specified quantity of water on the vehicle deck. This is considered important since, as already discussed in this paper, there have been a large number of recent ferry losses in Indonesia attributed to a loss of stability due to water ingress onto the vehicle deck without any damage to the ship being sustained.

LNA consider a closed vehicle deck design to be the safest option however should the vehicle deck be required to be opened to reduce the tonnage, the ship's ability to remain stable with water on the vehicle deck is important.

In order to perform the calculation, a NAPA stability model of the design was set at the service draught of 3.6 metres with a GM of 2.4 metres corresponding to a preliminary departure loading condition. A quantity of water was then added to the deck at the lowest freeboard point in 10cm increments. The residual stability was considered adequate if all SOLAS 90 deterministic stability criteria were met (this being considered equivalent to SOLAS 2009 probabilistic damage stability). The results indicated that even with 1.0m of water on the vehicle deck, all stability requirements are met. This is shown in Figure 7 below:

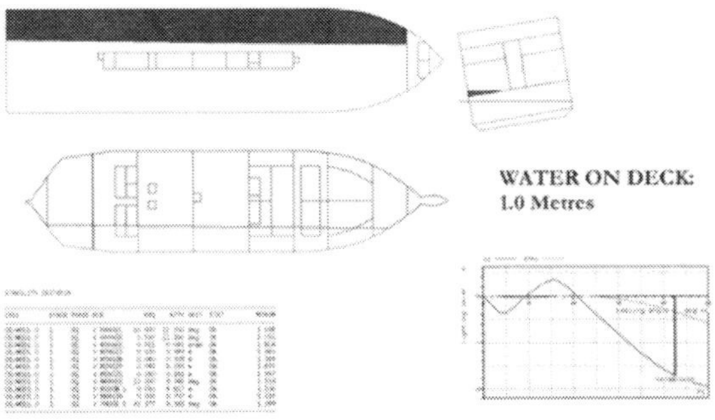

Figure 7: 1.0m Water on Deck SOLAS 90 compliance

This water on deck stability performance is vastly superior to existing vessels operating in the region as analysed by Anggoro [10]. This can be attributed to the relatively large beam of the LNA design together with the corresponding relatively large intact GM of the design. The heel angle with 1.0m of water on deck is 9.4°. Whilst this is a large heel angle for the vessel to adopt, it is based on the assumption that 1m of water has built up on the vehicle deck – an unlikely scenario if the vessel is fitted with blockage resistant scuppers and has a crew trained in damage control methods.

4. COMMERCIAL VIABILITY

4.1 Assessment Method

The commercial viability of the LNA design is, of course, an important consideration. In illustrating the commercial viability of the design, LNA has undertaken an analysis based on one vessel operating on the Balikpapan – Makassar Route, under the following assumptions:

- 340 Operational days per year
- Voyage distance 280nm
- Average Speed 15 knots
- One voyage per day (two-day return)
- Ship's generators used 265 days / yr
- Ship's main engines as required

In order to perform the analysis, data is needed in the following areas:

- Ship purchase cost
- Running costs
- Likely occupancy and freight rates

These will be considered in turn with the numbers brought together in the final section to determine the required occupancy rate for each voyage to achieve profitability.

4.2 Ship Purchase Cost

This is difficult to estimate without the benefit of quotes from Indonesian shipyards. LNA has undertaken a breakdown of the ship to assist shipyards in pricing the vessel. The lightweight of the vessel is preliminarily estimated to be 2228Tonnes of which there is:

- 1527 tonnes of steel
- 293 tonnes of passenger and crew space Outfit
- 47 tonnes of mooring equipment
- 39 tonnes of lifesaving equipment
- 220 tonnes of machinery
- 103 tonnes of piping and cabling

For the purposes of this analysis an estimate from similar Ro-Ro vessel projects from the ShipPax annual report "Designs" [11] was made. From 6 similar ships built in the period 2007 to 2011 it would seem that a value of US$20 million would be reasonable if the ship is build locally in Indonesia.

The possibilities for financing this amount are many but an allowance for this is considered in the running costs section which follows.

4.3 Running Costs

The fuel consumption of the vessel was estimated as part of the design process. From these consumption values the annual fuel cost can be estimated based on bunker prices from Bunkerworld [12]. This suggests an annual fuel cost of US$5.7 million.

Wages for the estimated 64 crew required to perform the 30 roles on the ship on a 24/7 basis were sourced from the Pelaut yahoo group [13]. This group features job advertisements for crew required for vessels operating in the Java Sea. An analysis of wages per month for the various roles leads to an annual wages cost of US$0.9 million.

Additional running expenses which include provisions, lubricants, spares and repairs and registration and port fees (estimated) amount to US$1.4 million. These were sourced for similar ships using the 'Opcost 2011 report' [13].

Finance costs were assumed for the entire US$20 million over 15 years at an average interest rate of 7%.

4.4 Freight Rates

The ship can carry a range of vehicles in varying numbers. For the purposes of the calculation an assumed cargo of the following has been assumed. This is the 100% occupancy benchmark.

- 670 Passengers
- 36 Motorcycles (2.40m each)
- 7 Private motorcars (4.80m each)
- 6 Pickups (4.80m each)
- 5 Small buses (7.00m each)
- 2 Large buses (9.50m each)
- 3 Small trucks (5.70m each)
- 2 Large trucks (7.00m each)
- 5 Trontons (9.50m each)
- 3 Trailers (12.45m each, trailer only)

Freight rates were determined using ASDP and PELNI published fare prices [5], [4] in terms of average freight rate for each vehicle/passenger type per nautical mile on similar routes. A small income from food and beverage sales of US$2 per person for the voyage was assumed. This gives an income of US$35,250 per voyage based on 100% occupancy.

4.5 Annual Balance

Considering the assumptions made earlier in this section, the likely purchase, finance and running costs and the achievable freight rates the percentage occupancy rate required for the first 15 years of the project to break even can be estimated. LNA has calculated this to be 85% occupancy average over 15 years.

Clearly this is not an attractive investment from a pure economics point of view. However LNA has calculated that if the finance was in the form of a 0% loan this rate of occupancy to break-even would fall to 78%. Similarly if half of the purchase price was gifted via an International or local economic aid program and the rest financed as per the original calculation (at 7% average), the breakeven point would be 76%. These values are more attractive to a potential investor, especially considering that the loan is over 15 years and the expected ship life could be 25-30 years the financial situation improves dramatically in the later years.

This form of International aid either in terms of gifts or interest free loans has many precedents as in the case of the building of MV Lady Samoa III [15]. This vessel along with two earlier built similarly sized vessels was granted to the Samoan government under the Japanese gift program in 2010.

5. CONCLUSION

This paper has examined the changing Indonesia ferry market from an independent consultant's point of view. Information on the ferry market is not readily available to those outside of the local industry however what information is available has allowed LNA to look at the trends and examine possible future developments.

It is clear that the industry is facing increased competition for passenger carriage in the form of air travel. Whilst a large percentage of the population still travel by sea, passenger journeys by sea are not keeping pace with domestic air travel. In order to increase revenue, LNA believes that a change from passenger only to Ro-Ro vessels is required. This trend has already begun with PELNI currently operating a pilot project vessel and ASDP operating a fleet of short sea crossings.

The LNA solution has been designed to operate on 150-300 nm routes which are longer than the majority currently operated by ASDP and similar to that offered by PELNI. LNA do not consider the Sunda Strait route to be a viable investment opportunity whilst the Sunda bridge solution is still under consideration.

The design presented is a 70.2m LBP vessel with a single vehicle deck and passenger day accommodation for 670. The design is to be of simple, inexpensive construction with savings made in the hullform, superstructure arrangement, loading / unloading arrangement and machinery installations. The relatively large beam and corresponding GM at sea has resulted in good stability during potential water on deck situations. Water on deck is known to have been the cause of several ship losses in the country. LNA consider that closing the vehicle deck weathertight and having such water on deck stability performance in the event of damage will result in this design having vastly improved safety over those vessels currently operating on similar routes.

Whilst utilising a new design such as the LNA concept will assist in growing the local economies of the regions it serves, LNA recognises that financing such a project will be challenging. An economic analysis undertaken by LNA for a Balikpapan to Makassar route has indicated that an occupancy rate of some 85% would be required to breakeven. This is likely to be too high to be attractive to potential business investors. If assistance were to be available however, possibly in the form of interest free loans or international aid programs this can be reduced to near to 75%. This is more viable, especially considering the finance is considered to be over 15 years whereas the life of the ship would be 25-30 years.

Such aid programs have been successful in the past; however they often result in the ship being built overseas. LNA believe that if the Indonesian government can assist in financing new vessels they can be built locally benefiting the local economy from day one.

Whilst the challenges of financing new Ro-Ro tonnage for Indonesia will be a challenge, such a new vessel(s) can help to grow trade between the islands of archipelago provide local jobs and improve passenger safety and comfort.

6. REFERENCES

1. 'Q1 Air Passengers up 15%, Tourism estimates double digit growth to Indonesia', Official Indonesian travel website www.indonesia.travel, 1st May 2012
2. 'Regional Forum on Domestic Ferry Safety, Bali, Indonesia 6 and 7 December 2011, Action Plan', www.imo.org, 16th December 2011
3. Ministry of Transportation & Ministry of Environment Republic of Indonesia, 'Indonesian Country Report on Environmentally Sustainable Transport Implementation', www.uncrd.or.jp/env/4th-regional-est.../08_BS2_Indonesia.pdf, 2008
4. PELNI Ferries, www.pelni.co.id, accessed 12th July 2012
5. ASDP Ferries, www.indonesiaferry.co.id, accessed 12th July 2012
6. LAM A., 'Connecting Java with Sumatra: An analysis of Indonesia's busiest ferry link', CRUISE & FERRY info FEB 12, February 2012
7. HUSSAIN Z., 'Indonesia Signs Deal for $10.9 Billion bridge Linking Sumatra to Java', Jakarta Globe, 21 April 2012
8. Badan Pusat Statistik, www.bps.go.id, accessed 13th July 2012
9. PriceWaterhouseCoopers, '2012 Outlook for the Retail and consumer Products Sector in Asia' page 33-34, www.pwc.com, accessed 12th July 2012
10. ANGGORO S., 'Analysis of the Intact Stability of Indonesian Small Open-deck Roll-on/Roll-off Passenger Ferries', University of New South Wales, March 2008
11. ShipPax, 'Designs 07-11' ISSN 1403-3437, 2007-2011
12. Bunkerworld, www.bunkerworld.com, Singapore 3 month average March-June 2012, accessed June 2012
13. Pelaut yahoo group, www.groups.yahoo.com/group/pelaut
14. Moore Stephens, 'Opcost 2011 report', Moore Stephens LLP, September 2011
15. Samoa Shipping, http://www.samoashipping.com/vesselinformation.htm

7. AUTHORS' BIOGRAPHY

Jennifer Knox is Managing Director of Lightning Naval Architecture (LNA) an independent design consultancy specialising in large commercial vessels. At LNA Jennifer works on technical analysis of passenger and dry cargo vessels and on the conversion and refurbishment of large passenger vessels. She holds bachelor degrees in Naval Architecture and in Commerce and has spent many months sailing the Indonesian archipelago.

Clive Michael Evans currently works as a Naval Architect at Lightning Naval Architecture in Sydney, Australia. A Masters graduate from the University of Southampton (UK), he is responsible for undertaking various consultancy projects within the organisation. He has particular experience in the new build and conversion of passenger Ro-Ro ships with a specialism in the intact and damage stability of these vessels.

*** Formatting Issue - Best Available Paper/Graphic ***

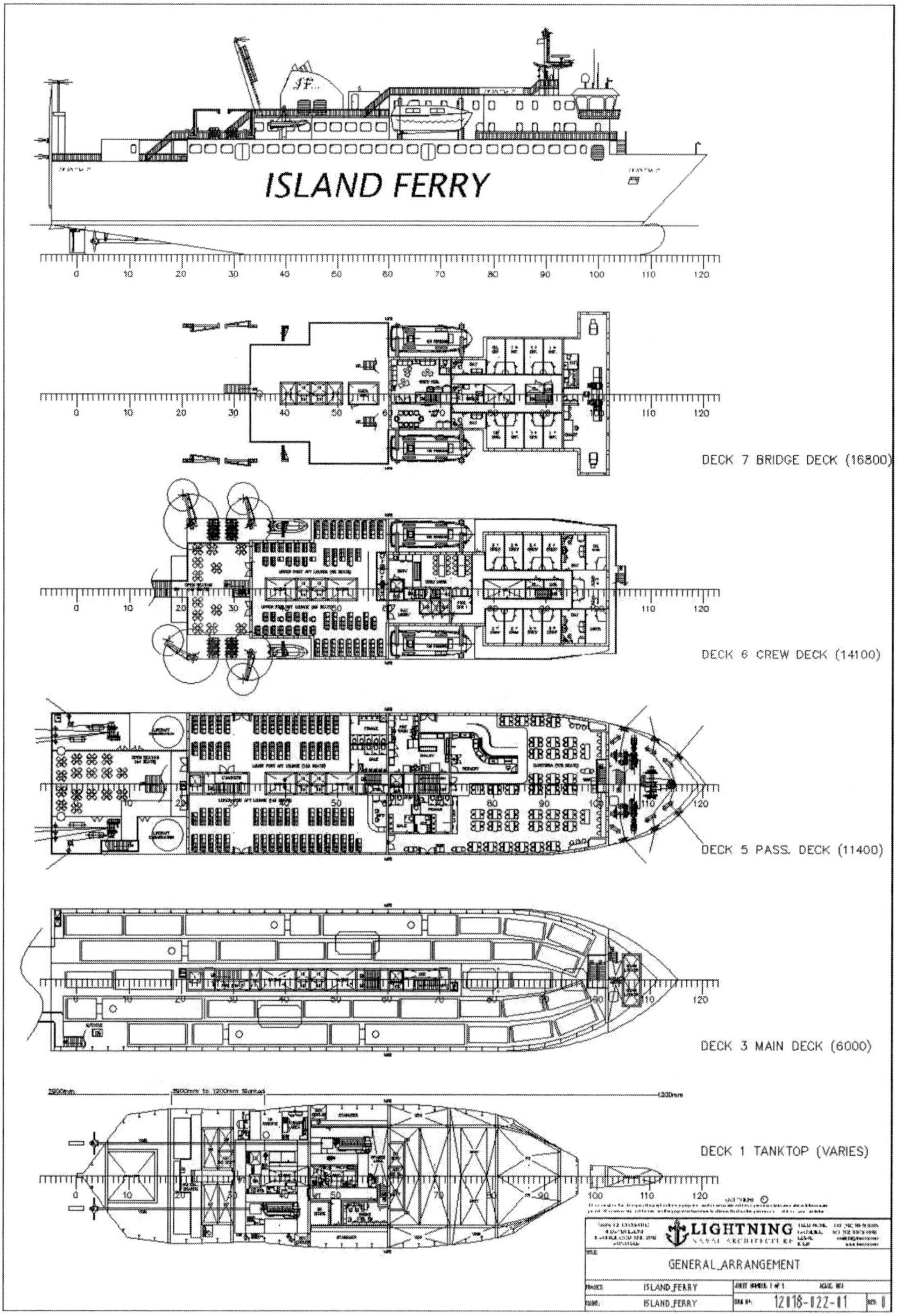

Figure 8: Concept design General Arrangement drawing

STABILITY OF SHIPS WITH LARGE BREADTH-DRAFT RATIO IN FOLLOWING AND QUARTERING SEAS

D. Paroka and **M. Djadir,** Hasanuddin University, Indonesia

SUMMARY

The International Maritime Organization (IMO) decided to revise its intact stability criteria from prescriptive based criteria to be the performance based one. The new criteria should be capable to investigate stability failure in seaways such as dead ship condition, pure loss of stability, parametric rolling and broaching. In case of pure loss of stability in following and quartering seas, delegation of Japan and USA proposed vulnerability criteria level one based on the hull form characteristics of ships. These proposed criteria have been tested by using several different hull forms. However similar characteristics with the Indonesian ro-ro ferries have never been used to validate the criteria. The main characteristics of Indonesian ro-ro ferries are small draught with large breadth or large ratio of breadth and draught. Therefore it is important to investigate the pure loss of stability in following and quartering seas in order to investigate effect of some variables.

This paper discusses about effects of wave direction, significant wave height and ship speed on roll motion characteristics of ships with large ratio of breadth and draught in irregular following and quartering seas. Two degree of freedom model is used to investigate the roll motion characteristics. The irregular wave is developed by using the ITTC wave spectrum and the linear and nonlinear damping coefficients of roll are estimated by using the Ikeda semi empirical method. The results show that the significant wave height has significant effect on the roll motion in following and quartering seas. Eventhought the ship may safely operate in the significant wave height of 2.25 meters with maximum roll angle of 0.545 radian or 31 degrees. In higher significant wave height, capsizing dangerous may occur due to pure loss of stability. The maximum roll angle does not significantly change due to alteration of the heading angle from wave direction. It means that the heading angle does not significant effect on roll motion in following and quartering seas. When the intial forward speed increases, the maximum roll angle decreases. However in a certain initial forward speed, the maximum roll angle increases because occurrence of resonance. This phenomena show that dangerous condition due to pure loss of stability may be avoided by change the ship speed.

NOMENCLATURE

A_{11}	Added mass in surge motion
A_{44}	Added inertia of roll motion
B_L	Linear damping coefficient of roll (s^{-1})
B_N	Nonlinear damping coefficient of roll (rad^{-1})
$B(x)$	Breadth of ship section (m)
D	Propeller diameter (m)
GZ_S	Righting arm in calm water (m)
GZ_W	Righting arm in wave (m)
J	Advance coefficient
$K_T(J)$	Trust coefficient
M_W	Roll excitation moment (N.m)
P	Propeller pitch
$R(u)$	Ship resistance (N)
$S(\omega,\chi)$	Wave spectrum
$T(n;u)$	Propeller trust (N)
V_A	Advance velocity (m s^{-1})
W	Ship weight (N)
X_W	Wave force in surge (N)
Z	Number of propeller blade
$d(x)$	Draught of ship section (m)
k	Wave number
n	Propeller revolution (rps)
t_p	Trust deduction factor
u	Surge velocity (m s^{-1})
ς_W	Wave amplitude (m)
ξ_G	Position of gravity in global axis (m)
ϕ	Roll angle (rad)
χ	Heading angle from wave direction (deg)
λ	Wave length (m)
ρ	Density of water (kg m^{-3})
ψ	Wave phase angle (rad)

1. INTRODUCTION

Since the International Maritime Oeganization (IMO) decided to revice its intact stability criteria, some researches regarding the intact stability problem in seaways have been conducted by many researchers. The main focus of the researches is ship stability in waves such as in following seas, quartering seas, beam seas and head seas. Several informations about capsizing prhenomena of ship due to ship stability in waves such as pure loss of stability in following and quartering seas, parametric rolling in following and head seas as well as roll resonance problem in bea mesas and surfriding following boaching in following and quartering seas have been obtained in last decade. In cases of pure loss of stability and parametric rolling, the dangerous condition arise due to large roll angle excited by alteration of restoring arm in waves. Based on these research results, the IMO decided to formulate the new generation of intact stability criteria in order to avoid possibility of dangerous condition due to those three phenomena. In other hand, the present intact stability criteria especially restoring arm characteristics namely area under the restoring arm curve up to certain heel angle, initial metacentric height, weather criteria as well as maximum heel angle due to heeling moments such as wind, passenger and turning moments.

The main structure of the new generation of intact stability criteria shown in Figure 1 has been agreed in the meeting of Intact Stability Correspondence Group (ISCG) developed by the IMO in 2011. As mentioned before that the general criteria and weather criteria in the present stability criteria still included in the new one. If the evaluated ship comply with both general and weather criteria then evaluation continued to dead ship condition, pure loss of stability, parametric rolling and surf-riding or broaching. Except the dead ship condition, the new criteria consists of vulnerability criteria level one, level two, direct assessment using numerical simulation and operation guidance. In case of dead ship condition, the group agreed to use capsizing probability approach under consideration that roll motion equation can be modelled as single degree of freedom and may be solved by linearization of the damping coefficient and restoring arm as propsed by Belenky [1], Bulian and Francescutto [2] and Paroka et. al. [3].

The new criteria for the last three dangerous conditions still under discussion because these phenomena depend to several variables and effect of its nonlinearity effect is significant. Therefore the evaluation methodology was separate into vulnerability level one, vulnerability level two, direct assessment and operation guidance. Some researches regarding the evaluation methodology for each level evaluation have been conducted by some researches.

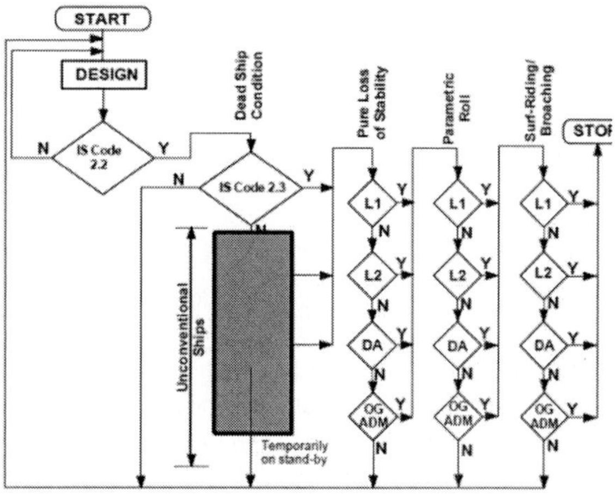

Gambar 1. Pendekatan evaluasi dari criteria stabilitas yang baru

In case of vulnerability criteria level one for pure loss of stability, Kubo, et. al. [4] proposed that dangerous condition for ship operating in following seas when the metacentric height estimated using the transverse moment of inertia of flat waterplan at the worst water level is negative. For the vulnerability criteria level two they propose the maximum restoring arm hydrostatically calculated with the worst sinusoidal wave surface. Belenky et. al. [5] proposed that the vulnerability criteria level one for pure loss of stability in following wave is described by the average of the vertical wall-sideness coefficient for the fore and after section of the ship. The dangerous condition may occur when the average vertical wall-sideness coefficient of the after body and the fore body of the ship is less than 0.75. The two proposals are actually similar because the nominal prismatic coefficient depends on waterline coefficient and regarding the minimum and maximum draught describe by the block coefficient of the ship. Regarding the vulnerability criteria level two for pure loss of stability in following waves, Bulian [6] proposed a probability approach under assumption that the roll damping coefficient is negligible small and applied the method to estimate probability of pure loss of stability of a container ship in irregular waves.

For ships with small draught or large ratio between breadth and draught, the possibility of negative metacentric height, significantly fluctuation of waterplan area and average of the vertical wall-sideness coefficient less than 0.75 when operating if following and quartering seas is very small. However, small alternation of restoring arm in waves may excite large roll angle due to initial heeling and very small damping coefficient of roll. Therefore a direct assessment for pure loss of stability in following and quartering seas for such ships is necessary in oreder to identify possibility of dangerous condition and effect of some variables on roll motion.

An example of ships with small draught and large ratio between breadth and draught is ro-ro ferry used as inter islands transportation in Indonesia. These ships have also quite small freeboard in order to easily operate the ship regarding the port condition. As a result, most of Indonesian ro-ro ferry could not comply with the IMO general criteria especially heel angle with maximum restoring arm less than 25 degrees [7] and [8]. However those ships still safely operate even several capsizing occur within the last decade. Generally the Indonesian ro-ro ferry may safely operate in Indonesian seaways in dead ship condition (beam seas) with capsizing probability less than the minimum accepted capsizing probability for public facility [8]. In order to ensure the safety of Indonesian ro-ro ferry in seaways, advance research regarding dangerous condition in seaways especially in following and quartering seas as well as head seas should be conducted. This is also aim to follow the new generation of intact stability criteria as proposed by the ISCG in Figure 1.

This paper discusses stability of Indonesian ro-ro ferry with large rasio of breadth and draught in following and quartering seas. This is important regarding developing of the new generation of intact stability criteria and validates the proposal vulnerability criteria level one for pure loss of stability submitted by the delegations of some countries in the ISCG. Effect of some variables on pure loss of stability in following and quartering seas such as wave height and direction relative to the ship and ship velocity are also investigate in this paper. The ship is assumed to have an initial heeling angle due to cargo shift in car deck. Several capsizing accident of Indoneian ro-ro ferry occurs due to cargo shift in the car deck because unproperly lashing of the car.

2. ROLL MOTION OF SHIP IN FOLLOWING WAVES

When a ship operates in following seas, wave excitation moment of roll is negligible small. However the ship may be in dangerous condition due to large roll angle especially if the the ship has small average of vertical wall-sideness coefficient or the ship has very small damping coefficient. The roll motion in following and quartering waves excited by the restoring variation depending on ship posisition relative to the wave. The minimum restoring arm occurs in hoggin condition in which the wave crest in the midship and the maximum restoring arm occurs when the wave crest in both the afterpeak and forepeak of the ship. If variation of restoring arm increases, the excited energy will increase. As a result, the roll angle due to restoring variation will also increase. If the ship body extrimly change in vertical direction then the restoring variation may be significant even the wave height is small. This is because the waterline area of the ship in hogging dan sugging condition will significantly change. Therefore some researchers proposed the waterplan area and the average vertical wall-sideness coefficient as the vulnerability criteria level one for the pure loss of stability in following and quartering waves [4] and [5].

Roll motion equation for ships operating in following and quartering seas may be modelled as two degree for freedom model coupled with the surge motion equation following [9]. These equation are developed using the ordinat system shown in Figure 2.

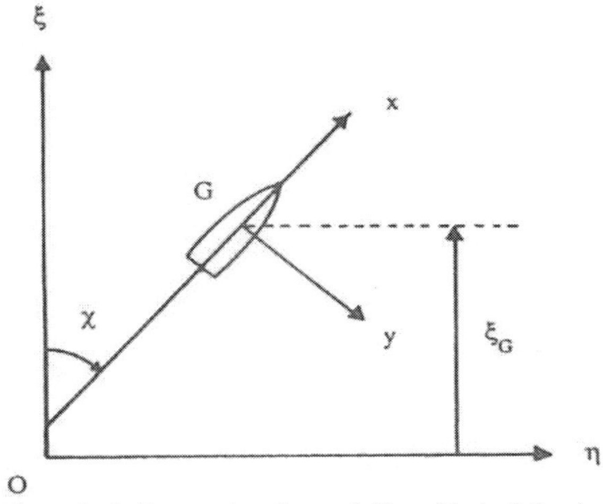

Figure 2. Ordinat system for modelling ship in following and quartering seas.

Here the surge motion equation is necessary in order to obtain the ship posisition in a certain time so that the ship posisition relative to the wave can be obtained for calculating the restoring arm. The two degree of freedom model can be written as follows.

$$(W + A_{11})\dot{u} - R(u) + T(n,u) = X_w(\xi_G/\lambda, \chi) \quad (1)$$

$$(I_{xx} + A_{44})\ddot{\phi} + B_L\dot{\phi} + B_N\dot{\phi} + W\big(GZ_s(\phi) + GZ_w(\phi,t)\big) = M_w(\chi,t) \quad (2)$$

The ship resistance as a function of surge velocity in the equation (1) are estimated by using the Holtrop Method for several surge velocities in calm water in order to obtain polynomial fungsi of the ship resistance as function of forward velocity.

The propeller trust as a function of propeller revolution and surge velocity is estimated using the following equation:

$$T(n,u) = (1 - t_p)\rho n^2 D^4 K_T(J) \quad (3)$$

where the thrust coefficient of propeller is estimated using the polynomial equation as shown in equation (4).

$$K_T(J) = \sum_{n=1}^{39} C_n(J)^{S_n}(P/_D)^{t_n}(A_E/_{A_0})^{u_n}(Z)^{v_n} \quad (4)$$

Value of each coefficient in the equation (4) is estimated based on statistical data of B Series propeller as a function of advance velocity [10]. Therefore, the coefficient thrust of propeller can be estimated for variation of ship velocity so that it can be modelled as polynomial equation in order to obtain propeller thrust in a certain ship velocity using the equation (3).

Surge wave force in the equation (1) may be estimated using the method developed by Umeda and Renilson [11] as follows:

$$X_w(\xi_G/\lambda,\chi) = \rho g \varsigma_w k \cos\chi \int_{AE}^{FE} C_1(x)\,S(x)e^{-\frac{kd(x)}{2}} \times \sin k(\xi_G + x\cos\chi)\,dx \quad (5)$$

Coefficient C_1 in this equation for each ship section may be estimated using the following equation [12]:

$$C_1(x) = \frac{\sin(k\ \sin\chi\,.B(x)/2)}{k\ \sin\chi\,.B(x)/2} \quad (6)$$

If the surge velocity has been estimated using the equation (1), the ship posisition relative to the wave can be estimated. Based on this ship position, the restoring arm in calm water and wave may be estimated, respectively. Finally the roll angle can be obtained by solving the equation (2).

3. SHIP DATA AND CALCULATION METHOD

The ship data use in this paper is a ro-ro ferry 600 GT operating as intern island transportation in Indonesia. This ship is desgned and duilt by national shipyard. The main dimension and ratio of main dimension as well as her body plan respectively are shown in Table 1 and Figure 3 as follow.

Table 1. Principle dimension of a ro-ro ferry 600 GT

Items		Values
Length between perpendicular (Lbp)	m	40.00
Breadth (B)	m	12.00
Draught (T)	m	2.15
Height (H)	m	3.20
Linear damping coefficient (B_L)	s^{-1}	0.008
Nonlinear damping coefficient (B_N)	rad^{-1}	0.0005
B/T	-	5.58
H/T	-	1.49
L/B	-	3.33
L/H	-	12.5

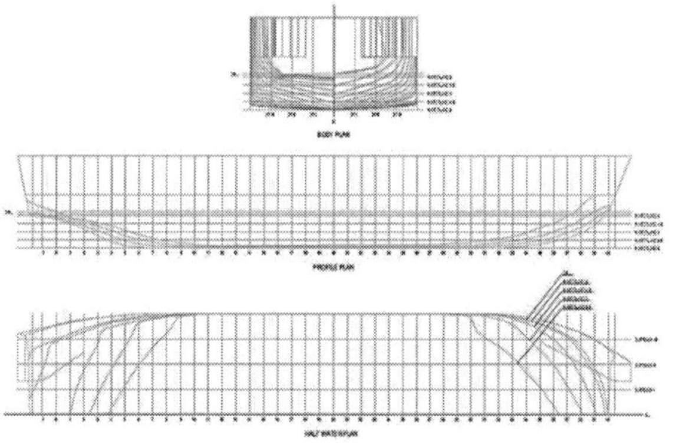

Figure 3. Body plan of subject ship

Characteristics hydrodynamics such as added inertia is estimated by using the strip theory. The roll damping coefficient both linear and nonlinear is estimated by using Ikeda semi empirical method. The surge wave force and the restoring arm in waves are estimated using irregular waves with wave length the same as the ship length. In order to model irregular waves, the ITTC wave spectrum is used and the effective wave profile is modelled by using the Grim effective wave [13]. The wave elevation based on the Grim effective wave can be written as the following equation:

$$\hat{\zeta}_{eff}(x,t) = a(t) + \zeta_{eff}(t)\cos\frac{2\pi}{L}x \quad (7)$$

dimana:

$$a(t) = \sqrt{2S\,(\omega,\chi)d\omega d\chi}\; F_A \cos(\omega t - kx\cos\chi + \psi)$$

$$\zeta_{eff} = \sqrt{2S\,(\omega,\chi)d\omega d\chi}\; F_C \cos(\omega t - kx\cos\chi + \psi)$$

$$F_A = \frac{\sin Q}{Q}$$

$$F_C = \frac{2Q\sin Q}{(\pi^2 - Q^2)}$$

$$Q = \frac{\omega^2\lambda}{2g\cos\chi}$$

In order to obtain effect of some variable on pure loss of stability in following and quartering waves such the wave height, the wave direction and ship velocity, numerical simulation in time domain is conducted with several different values of those variables. The ship velocity is determined by variated the propeller revolution with assumption that the power and the propeller characteristic are constant. This means that alteration of ship velocity only due to variation of the propeller revolution.

The range of ship velocity used for estimating the resistance is 6 – 15 knots. Based on the estimation results, polynomial equation of ship resistance as a function of ship velocity may be developed. Realtionship between the ship velocity and ship resistance is shown in Figure 4 and its polynomial equation shown in the equation (8).

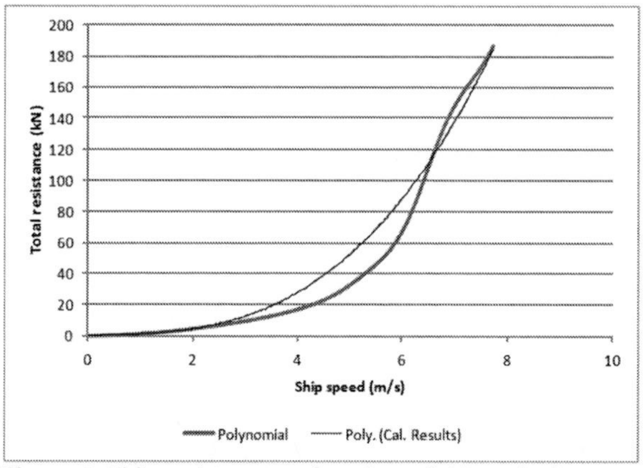

Figure 4. Ship resistance as function of ship velocity

$$R(u) = 0.3904u^3 + 0.6525u \quad (8)$$

Using the equation (4) for the same variation of ship velocity as the estimation of ship resistance, the thrust coefficient as function of advance velocity can be estimated. Here, the number of propeller blade, propeller diameter and aspect ratio of propeller are assumed to be constant. Variation of ship velocity occurs due to alteration of the propeller revolution. The obtained thrust coefficient for each advance velocity is shown in Figure 5. The polynomial equation of the thrust coefficient as function of advance velocity is shown in the equation (9).

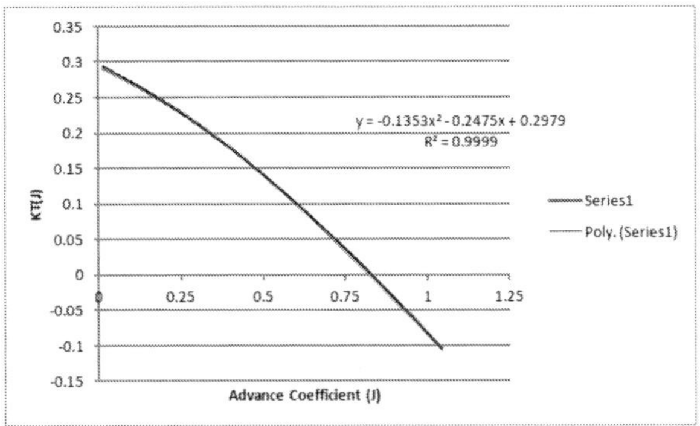

Figure 5. Thrust coefficient of propeller as function of advance velocity

$$K_T(J) = -0.1353J^2 - 0.2475J + 0.2979 \qquad (9)$$

where

$$J = \frac{V_A}{nD}$$

In order to obtain roll motion response, the equation (1) and (2) are solved in time domain using the second order Runga-Kutta method. The restoring arm in wave is calculated by considering the static trim with pressure due to both calm water and Froud-Krylov force. The subject ship is assumed to have an initial heel angle due to cargo shift on car deck. This condition is possible to occur in Indoensian ro-ro ferry escpecislly when the ship operates in rough weather.

In order to investigate effect of wave, the calculation in conduted for wave height form 1.50 meters up to 2.25 meters. The heading angle from wave direction is simulated from 0 degress up to 45 degrees with increasing of 15 degrees. The wave exciting moment for the heading angle geater than 0 degrees is neglected so that the roll motion only excited by the restoring arm variation in wave.

Effect of ship velocity on roll motion in following and quartering seas is investigated by changing the propeller revolution in order to obtain different ship velocity. Here three different propeller revolution are used which are propeller revolution for service speed, less than the propeller revolution for service speed and the last one is higher than the propeller revolution for service speed.

4. RESULTS AND DISCUSSION

4.1 EFFECT OF SIGNIFICANT WAVE HEIGHT

The roll motion for each significant wave height with heading angle nol degrees and the ship velocity the same as the service speed is shown in Figure 6. Here the simulation is conducted for time duration of 1500 seconds.

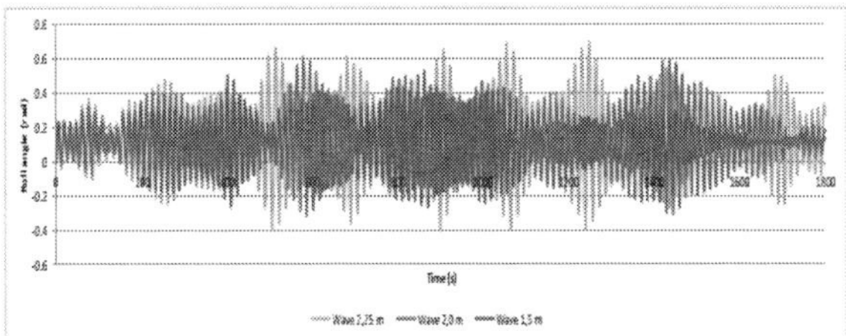

Figure 6. Roll motion for three different significant wave heights. The heading angle is 0 degress.

Figure 6 shows that the maximum amplitude of roll motion increases when the significant wave height increases. This is because the variation of restoring arm in wave increases when the significant wave height increases. Even the average of vertical wall-sideness coefficient is larger than 0.75 and the metacentric height in hogging condition for all significant wave heights is posisitve, the restoring arm variation may be significant due to significantly changes of ship body above the maximum draught especially both afterpeak and forepeak. Variation of restoring arm for significant wave height of 2.0 meters is shown in Figure 7. The alteration of the restoring arm variation for each significant wave height is smaller compared with the restoring variation of the container ship using by IMO as sample ship in the developing of the new generation intact stability criteria [9]. The large roll angle here may be also caused by small damping coefficieny of roll motion.

Within the duration of simulation, occurrence of capsizing is not identified in all significant wave heights and the roll motion is still stable. However the large roll angle mainly when the significant wave height is 2.25 meters, deck edge of the ship may immerced in seawater. This condition could be followed by another dangerous condition such as decreasing the ship stability due to trapped water on deck. As mentioned before that the Indonesian ro-ro ferries mostly have small freeboard, large roll angle should be considered as a dangerous condition even the roll motion still stable.

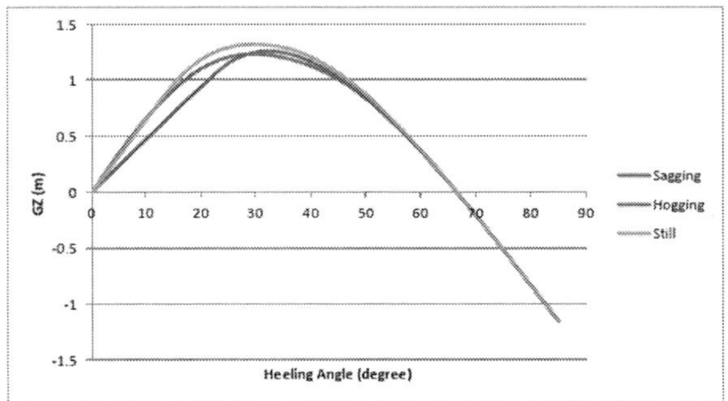

Figure 7. The restoring arm in calm water and wave. The significant wave height is 2.0 meters.

In the higher significant wave height for example 2.50 meters, dangerous capsizing occurs in which the roll motion become unstable or the pure loss of stability occurs after several second simulation. The time history of the capsizing is shown in Figure 8. Here the heading angle from wave direction is 0 degrees and the initial forward speed is 5.13 m/s. The roll angle increases with the exposure time and finally capzises.

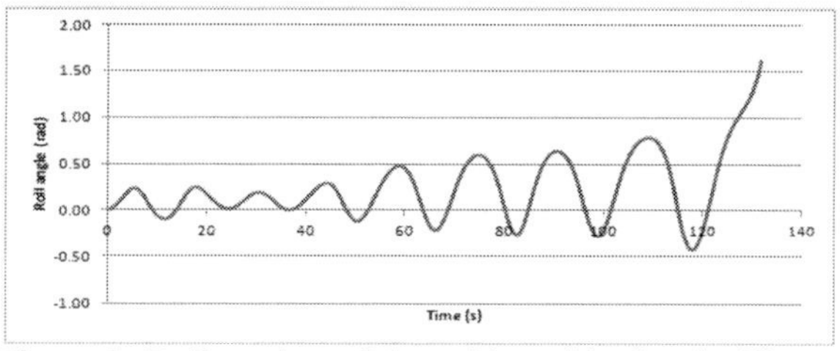

Figure 8. Roll motion of the subject ship for significant wave height of 3.00 meters. The heading angle is 0 degrees and initial forward speed is 5.13 m/s.

Regarding the vulnerability criteria level one for pure loss of stability in following and quartering wave proposed by delegations of some countries such as United States and Japan should be validated with different ships typology with the used topology when the criteria is developed. One variable should be considered in the new criteria is amplitude of restoring arm variation in wave as the factor induces roll motion in following and quartering seas. The dangerous condition may be identified when the difference of restoring arm in hogging and sugging condition excees a certain value which should be investigated in advance using several ship topology. This parameter may be combined with the roll damping coefficient as the one factor to avoid dangerous condition in following and quartering seas.

4.2 EFFECT OF HEADING ANGLE

Roll motion in following and quartering seas with heading angle from wave direction 0 degrees to 45 degreees with increasing of 15 degrees is shown in Figure 8. Here, the significant wave height is 2.25 meters and the time duration of simulation is 1500 seconds. The ship speed the same as the service speed of the subject ship. The sway motion and the wave exciting moment of roll are neglected. This means that the roll motion occurs purely due to the restoring arm variation in wave.

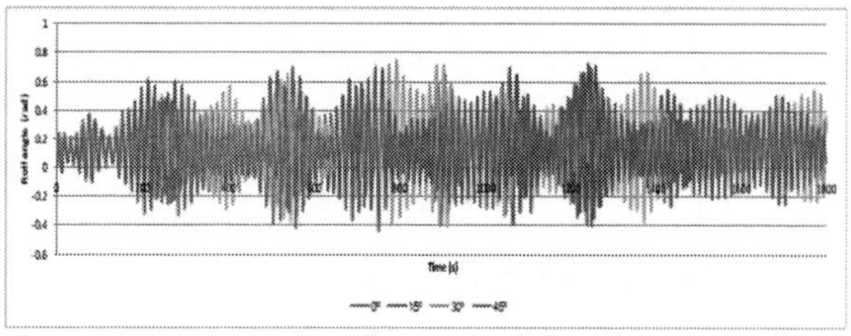

Figure 9. Roll motion for several heading angle from wave direction. The significant wave height is 2.25 meter.

The Figure 9 shows that maximum amplitude of roll motion is not significantly change due to variation of heading angle. This results show that alteration of heading angle from wave direction does not affect the restoring arm variation. Effect of the heading angle from wave direction may be identified when the effective wave length is larger than the ship length. It means that th effective length of wave for the heading angle will be the same as the ship length. The heading angle from wave direction affects only encounter frequency. Therefore the maximum amplitude of roll motion does not change but the maximum amplitude of roll occurs in different time for different heading angle from wave direction. In the initial time of simulation, the roll angle for all heading angle is relatively the same but different patern of roll motion arises in the next duration time of simulation.

Based on the above explanations, it can be concluded that the heading angle from wave direction has no effect on the amplitude of roll motion. In the real situation, the roll angle with larger heading angle is larger than that with smaller heading angle. This is because effect of wave exciting moment wich will increase by increasing the deading angle up to the heading angle of 90 degrees (beam seas condition). The resonance roll motion may be the other dangerous condition for the ship in this heading angle. In case of long wave, the surf-riding phenomena may occur followed by broaching. In order to investigate such dangerous condition, Umeda, et. al. [14] recommends a six degree of freedom mathematic model. Changing the ship direction relative to the wave is not recommended solution to avoid the pure loss of stability in following and quartering seas.

4.3 EFFECT OF SHIP SPEED

When a ship operates in following and quartering seas, the ship speed will affect the encounter frequency and the time spent by the ship in a position relative to the wave. The spent time in a relative position to the wave will directly influence the periode of restoring arm variation as the particular variables induce the roll motion.

Results of roll motion simulation for several initial forward speeds of the subject ship in significant wave height of 2.25 meters with the heading angle from wave direction of 0 degrees are shown in Figure 10.

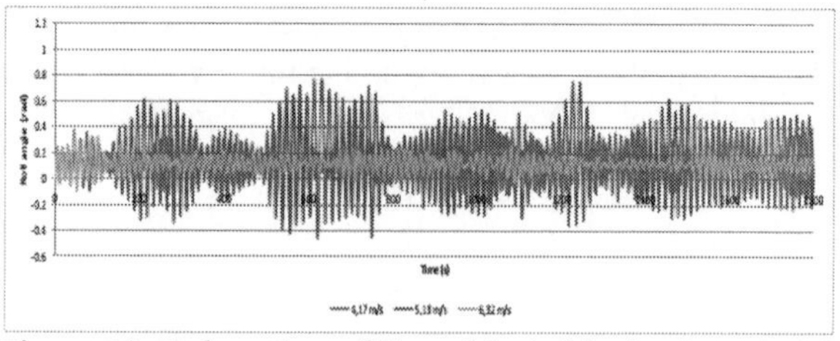

Figure 10. Rol motion of the subject ship for three intial forward speeds. The significant wave height is 2.25 meters and the heading angle is 0 degrees.

The amplitude of roll angle tends to decrease when the initial forward speed increases. However in an intial forward speed the amplitude increases as the initial forward speed increases. This peak amplitude occurs in several initial forward speeds but the value of amplitude decreases for higher initial forward speed. The alteration of roll motion amplitude due to variation of the initial forward speed is shown in Figure 11.

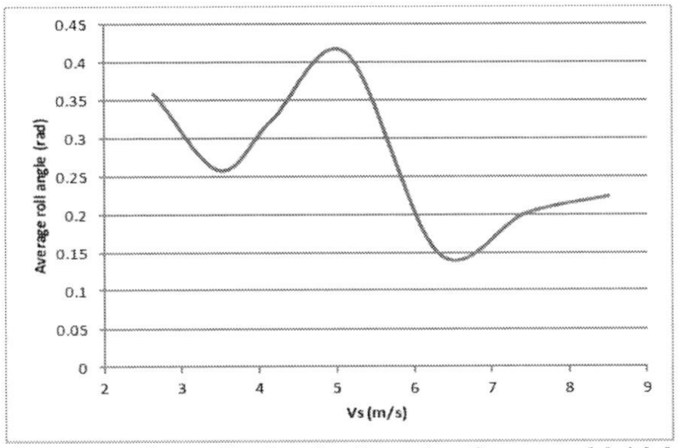

Figure 10. Th amplitude of roll motion for several initial forward speeds. The significant wave height is 2.25 meters and the heading angle is 0 degrees.

In a low intial forward speed, the wave encounter frequency is high. As a result, the period of restoring arm variation in wave also increases.Therefore the amplitude of roll motion tends to increase. By increasing the intial forward speed, the wave encounter frequency decreases and the period of restoring arm variation decreases. The ship will run with the wave for a longer duration of time. In the duration of time, the restoring arm tends to be constant. The amplitude of roll angle decreases because of contant restoring arm when the initial forward speed increases.

For a certain initial forward speed, the amplitude of roll angle increases due to increase of the initial forward speed. This is because of wave encounter frequency in such initial forward speed is the same as the natural frequency of roll. This means that the larger amplitude of roll angle due to resonance of roll motion. The resonance occurs when the wave encounter frequency is the same as multiple of the natural roll frequency. Eventhough, the resonance amplitude dercreases when the resonance frequency decreases or the initial forward speed increases. This phenomena is also caused by periode of restoring arm variation tends to be small in lower of resonance frequency. This fact shows that large roll angle or pure loss of stability in following and quartering seas can be avoided by change the ship speed. However, the other dangerous condition should be considered such as broaching phenomena in long wave cases.

5. KESIMPULAN

Based on the above calculation results and discussion, some conclusions can be remark as follows:

1. Capsizing dangerous due to pure loss of stability in following and quartering seas is not identified when the significant wave height smaller than 2.25 meters. However capsizing may occur if the significant wave height is larger than 2.25 meters. It means that the significant wave height significantly affects roll motion in following and quartering seas.

2. The maximum roll angle of the ship does not significantly change due to alteration of the heading angle from wave direction. It means that the heading angle form wave direction has no significant effect on roll motion in following and quartering seas. This fact shows that dangerous condition in following and quartering seas cannot be avoided by changing the heading angle from wave direction.

3. The initial forward speed has significant effect on roll motion in following and quartering seas because the periode of reastoring arm variation as the main factor affect the roll motion depends on the ship speed relative to the wave. The other reason is the ship may experience resonance roll motion when the ship speed induce encounter frequency the same as the natural frequency of roll. These facts shown that the large roll angle or dangerous of pure loss of stability in following and quartering seas can be avoid by changing the ship speed. Therefore vulnerability criteria of direct assessment and guidance operation are necessary in the new generation of intact stability criteria.

6. ACKNOWLEDGEMENTS

A part of this paper is results of research conducted by the authors during short visiting research in Osaka University of Japan supported by research grant from Project Implemetation Unit (PIU) of new campus of Faculty of Engineering Hasanuddin University. The authors sincerely thanks to the institution for it's supporting.

7. REFERENCES

1. Belenky, V.L., 'Piece-Wise Linear Method for the Probabilistic Stability Assessment for A Ship in Seaways', Journal of Ship Research, 1994.
2. Bulian, G. and Francescutto, A., 'A Simplified Modular Approach for the Prediction of the Roll Motion due to the Combined Action of Wind and Waves', Journal of Engineering for the Maritime Environment, 2004.
3. Paroka, D. and Umeda, N., 'Capsizing Probability Prediction for A Large Passenger Ship in Irregular Beam Wind and Waves', Journal of Ship Research, 2006.
4. Kubo, H. and Umeda, N., 'Designing New Generation Intact Stability Criteria on Pure Loss of

Stability on Wave Crest', the 4th International Maritiem Conference on Design for Safety, 2010.
5. Belenky, V., Bassler, C.C. and Spyrou, K.J., 'Development of Second Generation Intact Stability Criteria', Report of Hydromechanics Department, 2011.
6. Bulian, G., 'Checking Vulnerability to Pure Loss of Stability in Long Crested Following Waves: Aprobabilistic Approach', Journal of Ocean Engineering, 2010.
7. Paroka, D., 'Study on Stability Criteria for the Indonesian Ro-Ro Ferries (in Indonesia)', Report of Research Grand from Indonesian Directorate of Higher Education, 2009.
8. Ali, B., 'Evaluation of Bertin Coefficient on the Prediction of Roll Motion in Weather Criteria for Ships with Small Draft (in Indonesia)', National Seminar on Theory and Application of Ocean Engineering, 2011.
9. Umeda, N., Izawa, S., Sano, H., Kubo H. and Yamane, K., 'Validation Attempts on Draft New Generation intact Stability Criteria', the 12th International Ship Stability Workshop, 2011.
10. Carlton, J., 'Marine Propellers and Propulsion Second Edition', Elsevier Ltd., 2007
11. Umeda, N. and Renilson, M.R., 'Wave Forces on A Ship Running in Quartering Seas – A Simplified Calculation Method', 11th Australian Fluid Mechanics Conference, 1992.
12. Umeda, N. and Hashimoto, H., 'Qualitative Aspect of Nonlinear Ship Motions in Following and Quartering Seas with High Forward Velocity', Journal of Marine Science and Technology, 2002.
13. Grim, O., 'Beitrag zu dem Problem der Sicherheit des Schiffes im Seegang', Schiff und Hafen, Heft 6, 1961.
14. Umeda, N. and Yamakoshi Y., 'Probability of Ship Capsizing due to Pure Loss of Stability in Quartering Seas', Naval Architecture and Ocean Engineering of Japan, 1994.

8. AUTHORS' BIOGRAPHY

Daeng Paroka holds the current position of Head of Department of Naval Architect at Faculty of Engineering Hasanuddin University, Makassar. He is responsible for arrange the academic process and make planning for developing the department in the future. He holds PhD degree from Osaka University of Japan in 2007 in field of ship stability in waves. He has written some papers published in various international and national journals. He has also some experiences to attend in several international and international conferences such as Marine Technology Conference and national coenference such Seminar Nasional Teori dan Aplikasi Teknologi Kelautan (SENTA). Several researches in field of ship stability in waves have been also conducted especially in relation with the stability problem of Indinesian ro-ro passenger ferries.

Metamagfirul Djadir holds his undergraduate degree in field of Naval Architect on September 2012 in field of ship stability in following irregular waves. Apart of this paper is a result of his thesis.

SEAKEEPING ASPECTS IN THE DESIGN OF SURVEY VESSELS

B. Abeil, Maritime Research Institute Netherlands, The Netherlands

SUMMARY

The aim of the present paper is to provide a thorough overview of the most important seakeeping aspects related to survey vessels, in such a way that they can be integrated in future designs. The content of the paper is largely supported by recent hydrodynamic studies (numerical and experimental) performed at MARIN.

The first part of the paper deals with the ship motions, with a particular attention given to roll and roll stabilisation. The following part elaborates on crew performance. Aspects such as seasickness, interruptions of activities from excessive motions or accelerations are thoroughly discussed. Thirdly, the operability of sounding equipment in a seaway is addressed. The paper provides a description of the risk of bubble sweep-down and emergence of the underwater equipment and the related consequences on sonar downtime and potential damage. The paper concludes on a discussion on the water motions inside a moonpool, both at zero speed and in transit, and on possible ways to reduce them.

1. INTRODUCTION

The growing interest in scientific exploration combined with the ever increasing offshore activity has led to a surging demand for survey and research vessels. Nowadays naval architects are requested to design multi-purpose vessels fitted with highly-sensitive sonar equipments able to operate in hostile environments. Therefore, besides very good performance in calm water and excellent manoeuvring capabilities, today's designs must also meet stringent seakeeping requirements.

The present paper discusses in detail four of the principal seakeeping aspects that challenge the design of survey and research vessels. The work presented is the result of a broad literature study combined with the experience developed through discussions with designers and numerical and model test projects. In the first place, the ship motions as a result of wave excitation are discussed, with an emphasis on roll and roll stabilisation. In a second part, the focus is on the notion of crew performance and methods to investigate it. Thirdly, the consequences of the bow design on the risk of slamming and the operability of the sonar equipment placed very often under the keel are investigated. Particular focus is on bubble sweep-down and the related risk of sonar perturbation. In a fourth part, the paper deals with the water motion patterns inside moonpools, as these are expected to be more and more fitted onboard vessels to allow the launch and recovery of AUV's and ROV's.

2. SHIP MOTIONS

2.1 HEAVE AND PITCH

The heave and pitch motions are known to be linear in nature and are well predicted numerically (within 10 to 15 % accuracy in the worst cases). From heave and pitch the vertical motions at any location onboard the ship can be derived. The lowest vertical motions are usually found between Station 6 to 8, and the designers select this location to place for instance the moonpool or the A-frame for ROV launch and recovery.

2.2 ROLL

2.2(a) General overview

The natural roll period of most survey and research vessels is found to be within the range of wave periods that can be encountered worldwide. This, combined with a low roll damping, makes such vessels particularly exposed to large roll motions. Besides the strong sensation of discomfort experienced by the crew, large roll amplitudes and the transverse accelerations associated also generate excessive loads on the equipment. Large roll amplitudes are also responsible for an increase in thruster ventilation frequency, which can have severe implications for the manoeuvring ability of the vessel. Therefore, the behaviour of the vessel in roll should be examined with particular attention.

Figure 1 shows the contours of the Root Mean Square roll motion for a 98 m research vessel as function of the wave peak period (x-axis) and heading (y-axis), as calculated by 2D strip theory. 180 deg corresponds to head seas, and 0 deg to following seas. The significant wave height is 2 m and the speed 4 kn, which corresponds to a common sounding speed.

It is shown that the largest roll motion is obtained in stern-quartering seas for a wave peak period of 11.5 s. The RMS is then approximately 5 deg, which means that the largest roll amplitude is expected to be roughly 20 deg. The model tests performed showed quite good agreement with the numerical prediction.

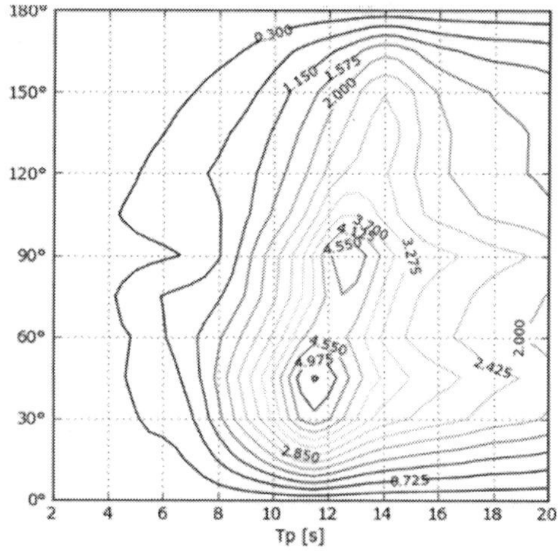

Figure 1: RMS roll motion of a 98 m research vessel, V_s=4kn, H_s=2m

2.2(b) Roll stabilisation

The usually strict requirements concerning the roll motion require the designer to adopt methods to reduce the roll motion.

Bilge keels are known to be a particularly effective way to reduce the roll amplitudes as they largely contribute to the roll damping at all speeds, including zero speed. It is in general noted that the length of the bilge keels is limited by the relatively small parallel midsection of survey vessels. Extending the bilge keels fore and aft of the parallel midsection is usually not a good option as they will tend to bend the streamlines, resulting in an increase of the calm water drag. In addition, the bilge keels ends will get closer to the waterline, which might result in bilge keel emergence and slamming when large vertical motions occur at the fore and aft ship. Also, the stabilising moment generated by the keels being a function of the third power of the distance to CoG, the contribution of the keel ends to the damping is expected to be substantially lower than that at the parallel midsection. Another possibility to increase bilge keel damping is to increase their height. Nevertheless, calculations and test results have shown that although a positive effect is noted on the extreme roll amplitudes, the influence on the roll RMS is clearly limited. As concluding remark the force acting on the bigger bilge keel are expected to increase substantially, which might lead to damage to the keels.

Besides bilge keels, common ways to reduce the roll motion consist of fin stabilisers and anti roll tanks. Fin stabilisers generate a stabilising moment either by creating a lift force at forward speed or by accelerating a considerable amount of fluid in a few seconds at zero speed. However, the fin stabilisers present the inconvenience of inducing additional resistance and require a substantial amount of power to work. Anti roll tanks (ARTs) are in that respect more interesting as they do not induce any additional resistance, can generate a considerable amount of roll damping at any speed, including zero speed and do not need much power.

ARTs work on the principle of having a volume of water travelling transversally in the vessel, in such a way that their motion is in counter-phase with the roll velocity. In this way, the mass of water induces a stabilising moment. There are two sorts of ARTs: free surface tanks, which consist basically of a hold that is more or less rectangular, filled with water, and the U-tube tanks, consisting of 2 side tanks connected by a duct at the tank bottom or close to the tank bottom. Advanced U-tube tanks feature an air duct system or are controlled by means of valves.

The free-surface tank has the main advantage of being relatively easy to design and install. The stabilisation of the vessel for several loading conditions is obtained by tuning the height of water inside the tank. However, the sloshing of water against the side walls of the ART yields noise and undesirable transverse force. U-tube tanks are less noisy and generate significantly less transverse forces, but they are more complex to design and they work in principle for one loading condition only as the natural period of the tank is mainly defined by the cross sectional area of the water duct [14]. Nevertheless, solutions have been found to adjust the performance of the tank for several loading conditions, for instance via the use of air-blocking valves.

Besides the tank layout and main dimensions, the performance of an ART is strongly related to the amount of internal damping present [7]. In figure 2 the transfer function of the unstabilised and stabilised roll motion derived from model tests is presented. On the left figure, an ART with low internal damping leads to an excellent reduction of the motion at resonance frequency, but yields adverse performance in shorter waves. On the right figure, a tank with a well tuned internal damping yields a good reduction of the motion at resonance without generating larger roll amplitudes in the longer and shorter waves.

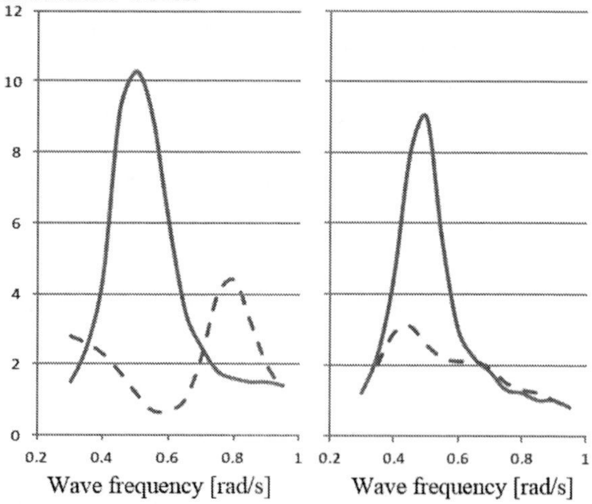

blue plain line: unstabilised; red dashed line: ART stabilised

Figure 2 – Roll stabilisation by means of ART on a 71 m survey vessel (left) and 98 m research vessel (right)

An increase in internal damping can be obtained by:

- setting obstacles in the water flow, for instance perforated plates;
- sharp corners in the water duct of U-tube tanks, resulting in increased vorticity;
- air-tight U-tank legs or reducing the flow of air from one leg to the other.

It should be emphasized that the prediction of the internal damping of a ART is currently the object of ongoing research at MARIN, notably through the use of RANS CFD. The results of the research will be published within the coming months.

2.3 GREEN WATER

The various and complex tasks conducted on the aft deck (e.g. the launch or recovery of an AUV) yield green water occurrences to have a strong impact on the vessel operability. A good prediction of the relative wave elevation can be obtained by means of a 3D frequency domain radiation-diffraction panel program and some comments can be already made on the risk of green water when the relative wave elevation is assumed to follow a Rayleigh distribution (a relation is then made between the RMS of the relative wave elevation and the probability of green water).

The relative wave elevation is also predicted during model tests by means of resistive wave probes.

3. CREW PERFORMANCE

Although the issue of crew and passenger discomfort and crew performance is usually mostly investigated for yacht, ferry and cruise vessel designs, the variety of the tasks that must be conducted onboard survey and research vessels by a limited crew size calls for more attention to the assessment of crew performance than paid until now. The aim of this section is to provide the designer with an up-to-date information on the various forms of discomfort that can affect the crew onboard and the methods that currently exist to estimate it.

As everybody may expect, the operability of a vessel is at maximum when all crew members are in good condition to carry out their tasks. Many studies conducted over the past 40 years show that relatively high ship motions lead generally the crew to experience seasickness (or in a more general way motion sickness) or generate more fatigue in relation with the additional energy expenditure to conduct their tasks.

Following the discussion in [8] and [14], the effect of ship motions on crew performance can be divided into:

- Motion sickness, in other words the sensation of discomfort induced by low-frequency ship motions (<1 Hz);
- Discomfort generated from high-frequency motions (> 1Hz);
- Motion induced fatigue; the energy spent by the crew to conduct a task is substantially increased due to the moving environment;
- Loss of balance: a given task is interrupted because the crew member lost balance, slid or fell.

3.1 MOTION SICKNESS

The origin of motion sickness is understood as a conflict between the information provided to the brain by the different sensory organs of the body (vestibular organs, eyes, somatosensory system). The associated symptoms are varied and can affect the physical and mental conditions of the crew, to name a few: stomach-ache, headache, nausea, vomiting, but also mental depression or considerable decrease of one's motivation.

Full scale experiments and model tests [10], [12], [17] have identified vertical motions as the main cause of motion sickness. However horizontal motions should not be disregarded. The experiments have also shown that not only the amplitude of the motion but also its frequency has an influence: people will mostly feel sick when the frequency of motion lies within 0.15 to 0.2 Hz [4].

As result of the research different models for motion sickness have been proposed. The most famous ones are the Motion Sickness Incidence (MSI), originally developed by O'Hanlon & McCauley [12] and the Motion Illness Rating (MIR or IR) by Bos [4]. The MSI was aimed as a direct measure of the proportion of people vomiting on a ship. However, the model was criticized for not taking the susceptibility of the population and the habituation to the motions into account and for being too radical in its description (people vomit or not, no steps in between). The MIR, developed later, proposes a more refined description of motion sickness with several steps. Also a model for a person susceptibility (e.g. women are more prone to feel sick than men) to motion sickness was proposed. The equation below shows the basic formulation of the MIR.

$$MIR = a_v \times h \times S$$

where a_v relates to the acceleration magnitude, h is an habituation factor and S describes the susceptibility of the population to motion sickness. A MIR calculation for a 71 m research vessel is provided in Figure 3. The criteria for MIR are discussed in section 3.5.

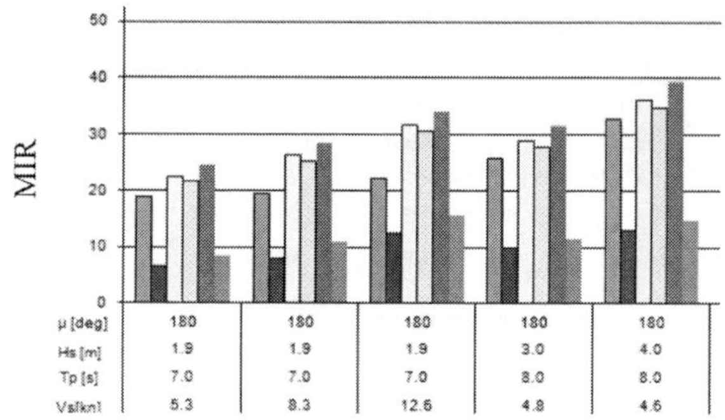

Figure 3 – Motion Illness Rating calculation for a 71 m research vessel, 1 hour exposure

3.2 HIGHER-FREQUENCY VIBRATIONS

The vibrations in the higher frequency range, caused by among others slamming, propeller, engine, machinery, were also identified as a source of discomfort and fatigue for the crew. Recent research has shown that the discomfort due to high-frequency vibrations is influenced by the amplitude and frequency of the vibrations as well as the duration of exposure. The Vibration Dose Value (VDV) model as found in the ISO 2631 [9] takes these parameters into account.

3.3 MOTION-INDUCED FATIGUE

It is commonly accepted that conducting a task is more difficult onboard a moving ship than on firm ground. The additional physical effort and attention required to cope with the ship motions when executing one's work, leads to a quick degradation of the mental and physical capability.

Much attention has been given to Motion Induced Fatigue (MIF). Experiments conducted in the TNO ship motion simulator [14] led to establish that motion induced fatigue could be related to the extra amount of oxygen consumed by one crew member (Time To Exhaustion model).

3.4 MOBILITY

Large ship motion can also cause the crew to lose occasionally their postural stability, in other words to fall, slide or for the least look for support to keep balance. The Motion Induced Interruption (MII) model was developed to predict the risk that one crew member will experience loss of stability. Criteria for the MII were presented in the NATO STANAG 4154 and by the US Coast Guard [14]. These criteria are given in the section below.

3.5 CRITERIA

Most of the research work conducted to determine the main factors and develop models for crew discomfort (motion sickness, fatigue) came to the conclusion that the notion of discomfort is highly subjective, and therefore difficult to qualify by means of criteria, and that the models (MSI, MIR, VDV, MII) should be mainly used as tools to compare several designs.

Nevertheless, some criteria were presented. The NATO STANAG 4154 states that a MSI of 20% in 4 hour exposure can be considered as acceptable, while the USCG gives a recommended 5% in 30 minute exposure. At MARIN use is mainly made of the more recent MIR index, and the value of 10 is applied as a threshold for most studies related to yachts. Considering that the crew of a survey vessel can be considered as more adapted to motions, the criterion might be relaxed.

For the VDV index a maximum of 8.5 based on 30 minute exposure is presented as a reason for voluntary speed reduction on a patrol vessel. Stricter criteria are proposed for cruise vessels and ferries (0.5 to 1), which indicate that a suitable criterion for survey and research vessels should lie in the middle. According to Griffin, relayed in Table 3 presented in ISO 6954, a VDV value of 3.2 measured in a workshop will yield probable adverse comments.

Considering the MII, the NATO STANAG presents a criterion of 1 occurrence per minute (MII=1) as an acceptable level. The USCG provide a more relaxed criteria, i.e. MII=2.1 tips per minute.

Finally, Nordforsk [11] provides a list of criteria related to the RMS vertical acceleration. In the list each criterion value is associated with a description. For instance, a value of 0.1 g is associated with the maximum acceleration for which people not so well adapted (the example of scientific personnel working on a research vessel is given) can still perform intellectual work.

4. DESIGN OF SONAR DOMES

4.1 GENERAL ASPECTS

Chakravorty and & Bhujanga Rao [2] identify machinery, propeller and flow as the main sources of noise that can affect the functioning of sonar domes located underneath the keel. Because machinery-generated noise and propeller noise are transmitted via the keel plate, the sonar equipment is usually located as much forward of the vessel as practically possible. However, they become then more vulnerable to the air bubbles present in the flow at the bow (bubble sweep-down) and to large impact pressures that can result from emergence.

4.2 BUBBLE SWEEP DOWN

Bubbles present in the flow area near the bow are believed to have four origins:

- bubbles that are naturally present in the flow around free-surface, created by breaking waves;
- keel emergence at the bow yielding a pocket of air to be caught below the keel during re-entry into water;
- bow thrusters emit bubbles in the flow after emerging;
- wave run-up at the bow when in transit can cause spray near the stem, creating bubbles in the free-surface.

In transit condition the bubbles are transported with the flow. Depending on the orientation of the streamlines some of the bubbles may come in contact with the sensors. Today CFD tools are able to predict the orientation of the streamlines in the steady flow for various speeds and therefore this issue can be easily solved.

However, when the vessel evolves in a seaway the motions at the bow yield a dynamic variation of the streamline orientation, so that the sonar equipment may no-longer be safe from exposure to bubbles. Consequently, another approach for the estimation of bubble sweep-down and the effect on sonar position is required.

At MARIN a combined numerical and model test approach is proposed for the assessment of bubble sweep-down. During the numerical phase a preliminary estimate of the risk of keel and bow-thruster emergence in various wave and sailing conditions is made. Calculations are conducted using MARIN's 3D potential code PRECAL. The results allow to draw first conclusions on the risk that bubbles will be entrapped by the keel or in the bow thrusters after emerging. The calculations are followed by model tests in waves during which so-called "underwater flow observations" are performed.

Underwater flow observations consist in the observation of the streamlines at the bow by means of an underwater camera. Because the forces acting on a bubble being also subject to scale effects, the direct observation of bubbles released in the flow is not an option. By releasing coloured water at specific locations the streamlines become visible. The presence of coloured liquid around the sensor location is controlled and quantified visually. A snapshot of such underwater observation is presented in Figure 5. In Figure 5, streamlines, marked by white arrows, were followed: two at the stem at different vertical positions close to the waterline and one location directly behind the bow thruster tunnel.

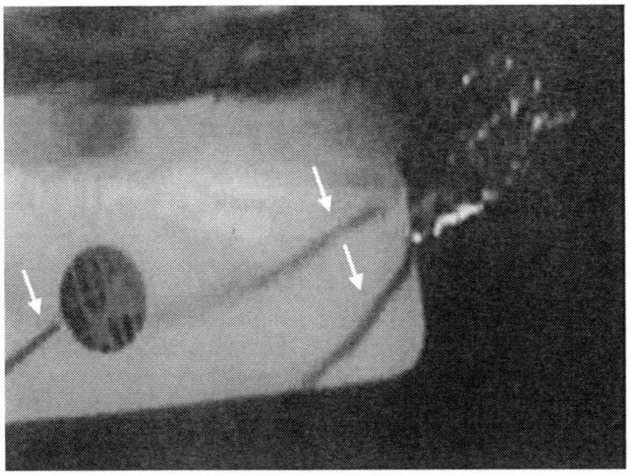

Figure 5 – Underwater flow observations during seakeeping tests of a 71 m survey vessel

The analysis of the video-recordings allows to estimate the frequency at which one of the marked streamlines reaches the sonar equipment. Also the flow patterns around the bow for various wave conditions (height, period) and sailing conditions can be determined. However it was found difficult to conclude on the risk that the bubbles will affect the sonar as the density of the bubbles in the streamlines remains unknown.

4.3 SONAR EMERGENCE AND SLAMMING

Besides the issues caused by the presence of bubbles in the neighbourhood of the sensor area, the flat, horizontal surface of the forward part of the keel or the gondola accommodating the sensors is likely to experience slamming when the motions at the bow yield emergence of the sensors. Consequently the risk of keel and gondola emergence should be carefully investigated. If preliminary calculations or model tests indicate that the keel or the gondola will experience slamming, the impact pressures on the gondola should then be estimated and compared with the maximum pressures that can be taken by the sensors. During the model tests the relative wave elevation around the sensor area is directly measured by means of resistive wave probes. Sometimes the impact pressures below the keel on the gondola plate are measured by means of pressure transducers. The amplitudes of the relative wave elevation are compared with the relative position of the keel or gondola and the probability of emergence is determined.

Although no criterion exists for the emergence of gondolas, recent seakeeping studies have shown rates of emergence of a few percents (less than 5%) of encountered waves. The measurement of the slamming pressures should undergo further post-processing to determine the most probable maximum value that can be compared with the robustness of the sensor accommodation.

5. MOONPOOL WATER MOTIONS

Although moonpools are usually found on drillships, cable-laying vessels, rock-dumping vessels, diving support vessels and production barges, they are also present on several survey and research vessels. On survey vessels moonpools are used for the launch and recovery of Remotely Operated Vessels (ROV's).

The experience developed in moonpool operation has shown that the activities inside the moonpools need to be interrupted when the motions of water inside become too large. Water motions inside the moonpool may occur in the following conditions:

- in transit in calm water;
- moored in waves;
- in transit in waves.

5.1 ORIGIN AND NATURE OF MOTIONS

The oscillation of the water column in calm water was found to be caused by the development of vortices that start at the upstream submerged moonpool edge, as illustrated in Figure 6, which will then rotate inside the moonpool. Depending on the advance speed the oscillation can enter into resonance, resulting in considerably large amplitudes.

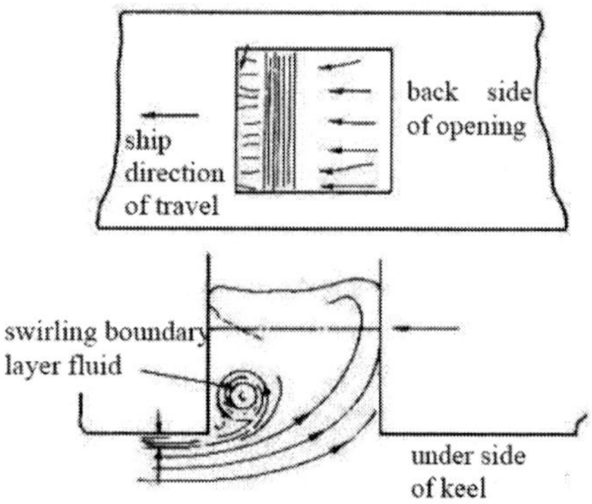

Figure 6 – Water motions in a rectangular moonpool [4] [5]

In the case of a vessel at zero speed in waves, the motions are generated by the combined action of the incident wave, wave diffraction effects and the ship motions ([1], [6], [13]). As for the calm water condition, resonant motion can be observed when the frequency of the excitation comes close to the natural frequency of the moonpool.

Observations during model tests and at full scale ([3],[6],[16]) show two main patterns of motions:

- a pure vertical motion of the water column, also referred to as "piston" mode;
- a longitudinal sloshing mode.

The motion pattern that will be observed depends on several parameters. Model tests performed in calm water [16] show that the geometry of the moonpool will have a major influence: squared moonpools are more likely to show piston motions, while monpools with a L/B ratio higher than 1 will show increasing dominance of the sloshing mode. The frequency of the excitation plays also a role as the sloshing mode is noted to be higher than the piston mode [16]. Formulations for the natural frequencies of the piston mode and sloshing mode have been proposed by Fukuda and Molin [16]. The formula of the natural frequency of the piston mode according to Fukuda is provided below:

$$\omega_n = \sqrt{\frac{g}{d+0.41\sqrt{S}}}$$

where d is the draught and S the moonpool plane area.

5.2 REDUCING MOONPOOL MOTIONS

Means to reduce the motions inside the moonpool in transit in calm water have been widely investigated. Some solutions, such as wedges, grids or flaps aim at preventing the upstream vortex from appearing. Some other target the motion itself: horizontal flanges (with flaps), damping chambers or a top cover are the most common examples investigated. Most of them turn out to be effective during model tests. However some present the major drawback of obstructing partially or completely the moonpool. This means that they must be movable in order to be able to conduct the necessary operations, a need for mobility that is usually not desirable. In addition, the need for regular maintenance might balance substantially the added value of the motion damping devices.

5.3 INVESTIGATION OF MOONPOOL MOTIONS AT MARIN

5.3(a) – Model tests

At MARIN the prediction of moonpool motions is mostly conducted through model tests. The oscillation of the water column is recorded at several locations inside the moonpool so that the different modes of motion (pure vertical oscillation and sloshing motion) can be distinguished. A comparison of the amplitudes measured in the moonpool with the deck height allows to conclude on the risk of green water.

5.3(b) – Moonpool Joint Industry Project (JIP)

MARIN is currently starting a new Joint Industry Project (JIP) called Moonpool JIP. The objective of this project is to study excitation mechanisms and scale effects inside moonpools. To reach these goals use will be made of new techniques, including the Particle Image Velocimetry for the visualisation of the flow inside the moonpool in transit in calm water, and a more pronounced use of numerical means for the prediction of moonpool motions in waves.

6. CONCLUSIONS

In this paper some of the main seakeeping aspects that should be considered in the design of a survey and research vessel were presented.

The high probability for the future ship to experience resonant roll in beam and stern-quartering seas means that a particular attention should be given to roll damping. Besides bilge keels, anti-roll tanks appear to be the most suitable solution for roll stabilisation as they can provide high motion damping at all speeds. The paper has shown that the amount of internal damping of the tank should be well considered as too little damping might lead the tank to act as excitatory in shorter or longer waves.

Although investigations related to crew performance and passenger comfort have been up to now more or less confined to passenger vessel and yacht design, the variety of tasks and the limited number of crew available on a survey vessel implies that more attention should be given to crew performance in the future designs. The paper described how crew performance is currently being investigated and suggests to apply the methods to the future design of survey and research vessels.

The high sensitivity of the underwater sonar equipment to environmental noise and bubbles implies that a thorough analysis should be conducted to determine the risk of downtime due to excessive noise. MARIN suggests an approach to investigate the path of bubbles where combined numerical means and model tests featuring underwater flow observations are employed.
In the final chapter, the issue of moonpools was tackled. The paper provided some background information on the origin and nature of motions of the water column inside the moonpool, as well as a discussion on ways to reduce or annihilate them. At last the new Moonpool JIP, which will allow better insight in the moonpool motions, was mentioned.

7. REFERENCES

1. ALBERS, A. 'The water motions in a moonpool', Ocean Engineering, Vol 11(6), pp. 557-579, 1984
2. CHAKRAVORTY, P.K. , BHUJANGA RAO, V., 'Some aspects in the Design of Sonar Domes', Def Sci J, 32(1) pp 47-54, 1982
3. COTELEER, A., 'Moonpool water motions caused by forward ship speed', Master Thesis, 1999
4. DALLINGA, R.P., BOS, J.E., 'Cruise ship seakeeping and passenger comfort', Proceedings of the Human Performance at Sea (HPAS) Conference, 2010
5. ENGLISH, J.W., 'A means of reducing the oscillations in drillwells caused by the vessel's forward speed", The Naval Architect, Vol. 3, pp. 88-90, 1976
6. GAILLARDE, G., COTELEER, A., 'Water motion in moonpools - Empirical and numerical approach', 2005
7. GOODRICH, G.J., 'Development and design of passive roll stabilisers', Transactions of Royal Institution of Naval Architects, 111, 81, 1969
8. GRIFFIN, M.J., 'Effects of vibration on people', Handbook of Noise and Vibration Control, 2007
9. INTERNATIONAL STANDARD ISO 2631-1 'Mechanical vibration and shock – Evaluation of human exposure to whole body vibration, 1997
10. LAUWTHER, A., GRIFFIN, M.J., 'A survey of the occurrence of motion sickness amongst passengers at sea', Aviation, Space and Environmental Medicine, 59(5), pp 399-406, 1988
11. NORDIC CO-OPERATIVE PROJECT, 'Assessment of ship performance in a seaway', 1987
12. O'HANLON, J.F., McCAULEY, M.E., 'Motion Sickness Incidence as a Function of the Frequency and Acceleration of Vertical Sinusoidal Motion', Aerospace Medicine, 45(4) pp. 366-9, 1974
13. MATUZIAK, J., 'Water column motion in a moonpool of a ship', Rakenteiden Mekniika, 30(2) pp 75-87, 1997
14. STEVENS, S.C., PARSONS, M.G., 'Effects of Motion at Sea on Crew Performance: A Survey', Marine Technology, 39(1) pp. 29-47, 2002
15. STIGTER, C., 'The performance of U-tanks as a passive anti-rolling device', TNO Report No. 81 S, 1966
16. VEER, R. v., THOLEN, H.J., 'Added resistance of Moonpools in calm water', Proceedings of the 27th OMAE Conference, 2008
17. WERTHEIM, A.H., BOS, J.E., BLES, W., 'Contributions of roll and pitch to sea sickness', Brain Research Bulletin, 47(5), pp. 517-524, 1998

7. AUTHOR'S BIOGRAPHY

Bastien Abeil holds the position of Project Manager within the Department Ships of the Maritime Research Institute of The Netherlands (MARIN). He received a Master of Science degree in Naval Architecture and Ocean Engineering from the Chalmers University of Technology in Sweden and from the ENSTA Bretagne Engineering School in France in 2007. He is responsible for the execution of contractual and research projects related to the seakeeping of ships through numerical simulations and model tests.

RUDDER AREA EFFECT ON HEELING MANOEUVRE OF A FERRY

N. Yuniarsih, A.H Muhammad and **M.A Djabbar,** Hasanuddin University, Indonesia

SUMMARY

Ferry boat operation, particularly in Indonesia is increasing widely. Research on the boat by naval architects is intensifying as well.

The aim of the study was to determine the effect of rudder on ship heeling manoeuvre.
Matrix Laboratory (MATLAB) simulation was used to predict the heeling angles. The simulation utilized model based on the concept of Mathematical Modelling Group (MMG). The model includes theory, separating components of hull equations, propeller and rudder as well as their interaction (hull, propeller, and rudder). The simulation results are compared with sea trial performance of ferry of *KMP Sultan MURHUM.*

It was obtained that the heel angle of 3.10 , 3.15, 3.20 degrees for 1.632, 1,819, 2,078 square meter rudder area respectively.

The results indicate that the greater the rudder area the greater the heel angle during the manoeuvre.

NOMENCLATURE

ϕ	Roll angle
ρ	Density of water
β	Drift angle
δ	Rudder angle
ψ	Course angle or heading
Λ	Rudder aspect ratio
ε	Phase angle
a_H	Rudder to hull interaction coefficient
A_D	Advances diameter
A_R	Rudder area
B	Ship breadth
B_{44}	Added inertia of roll motion
C_1, C_2, C_3	Constants for open water propeller
CB (*Cb*)	Block coefficient
D_P	Propeller diameter
D_T	Tactical diameter
f_a	Gradient of the lift coefficient of rudder
F_N	Normal force acting on rudder
G	Center of gravity of vessel
I_{ZZ}	Moment of inertia with respect to the z-axis
J	Advance coefficient
J_{ZZ}	Added moment of inertia around z-axis
K	Moment with respect to the x- axis
K_Q	Torque coefficient
K_R	Momen of rudder
K_T	Thrust coefficient
LOA	Ship length overall
LBP	Length of between perpendiculars
m	Mass of ship
n (rpm)	Propeller revolution
V_A	Advance velocity (m s^{-1})
N	Moment with respect to the z- axis
P	Propeller pitch
P/D_P	Propeller pitch ratio
t_p	Trust deduction factor
T	Draught of vessel
r	Turning rate or angular velocity
r'	Non-dimensional turning rate
u	Velocity in x- direction (surge)
$\dot{u}$	Acceleration in x-direction (surge)
U	Vessel velocity
U_R	Rudder inflow velocity
v	Velocity in y-direction (sway)
$\dot{v}$	Acceleration in y-direction (sway)
w_p	Effective wake fraction coefficient at propeller
w_{p0}	Effective wake fraction coefficient of propeller in straight running
x, y, z	Ship-fixed co-ordinate system
x_0, y_0, z_0	Space-fixed co-ordinate system
x_H	The distance between the centre of gravity of hull ship
x_G	Centre of gravity (positive if forward of amidships)
x_R	The distance between the centre of gravity of ship and centre of rudder lateral force
X	Force in x-direction acting on ship
X_0, Y_0	Forces in the x_0- or y_0- direction
X_P	Propeller thrust
X_R	Rudder force in x -direction
Y	Force in y-direction
Y_R	Rudder force in y -direction
Z	Number of propeller blades

1. INTRODUCTION

The number of ferry (boat) carrying trucks, bus, car, motor cycle, passengers, etc. is increasing in Indonesia. The operation could improve national economic growth. In ferry building, from design to delivery, a number of test must be conducted, particularly the ability of ship operations. The test includes testing the ship speed and maneuverability.

Maneuvering characteristics are very important criteria to make sure ship can operate in certain situation and location. Maneuverability is very critical aspect especially in harbour and offshore operations such as tow and tug to avoid collision. Noor [1] suggested that maneuverability is an important characteristic that must be predicted during the early design of ship. The simulation of ship maneuvering has now being progressed quite well and is very useful in designing the ship [2].

Ship maneuverability is also directly related to navigation safety and economy. In some restricted waters, marine accidents can happen if the ability to maneuvre the ship is not adequate. On the other hand, the dynamic stability of ships with poor course could only maintain its position by using a control device as often as possible. The consequences for a ship like this is not just a voyage being longer than planned, but also more energy is consumed by the control device [3].

Many accidents are caused by poor ship maneuvering quality. General consideration of it is not a major problem, resulted some boats that are built have the low quality of manoeuvering [4].

Ship designers need to consider the existing rules, both nationally and internationally. In 2002, the International Maritime Organization (IMO) [5] issued a Resolution MSC.137 (76) "Standards for Shi p Maneuverability", to develop safety standards and to ensure the safety of cruise ship operating at sea.

In principle, manoeuverability is greatly influenced by hull design, propulsion, rudder and steering systems. A number of these elements is in a direct significant influence on the hydrodynamic force and moment when ship manoeuvres [6].

2. METHODOLOGY

2.1 MATHEMATICAL MODEL

In analyzing the ship maneuvering through computer simulation, mathematical modelling is important to be developed including hydrodynamic coefficients derived here. In this study, a mathematical model based on the equations of motion, Eq. 1 of the ship developed (4 degrees of freedom) the surge, sway, yaw and roll motions.

The mathematical model for maneuvering motion can be described by Eq. 1 using the coordinate system, Fig.1..

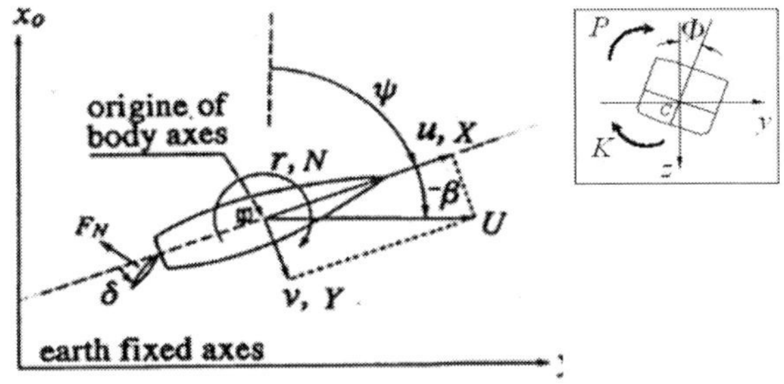

Fig 1. Ship Coordinate System

$$X = m(\dot{u} - rv)$$
$$Y = m(\dot{v} - ru) \quad (1)$$
$$N = I_{ZZ}\dot{\psi}$$
$$K = I_{XX}\dot{p}$$

The notation of u, v and r are velocity components at centre of gravity of ship (G). U represents resultant of the speed of the ship . X, Y and N represent the hdrodynamic forces and moment acting on the mid-ship of hull. These forces and moments can be defined separately into the different elements of physical force and moment of the ship in accordance with the concept developed by Ogawa and Kansai [7], Eq.2.

$$X = X_H + X_R + X_P$$
$$Y = Y_H + Y_R + Y_P$$
$$N = N_H + N_R + N_P$$
$$K = K_H + K_R + K_P$$
$$(2)$$

Where, the subscript H, P and R refer to hull, propeller and rudder. Force and moment induced by hull (X_H, Y_H, N_H and K_H) in principle is an approximation of polynomiual regression β and r'. Furthermore the coefficients of these equation can be termed as derivative of the hydrodynamic coefficients. The equation can be expressed by Eq. 3 :

$$X_H = \tfrac{1}{2}\rho L d U^2 (X'_0 + X'_{\beta\beta}\beta^2 + (X'_{\beta r} - m'_y)\beta r' + X'_{rr}r'^2 + X'_{\beta\beta\beta\beta}\beta^4)$$
$$Y_H = \tfrac{1}{2}\rho L d U^2 (Y'_\beta\beta + (Y'_r - m'_x)r' + Y'_{\beta\beta\beta}\beta^3 + Y'_{\beta\beta r}\beta^2 r' + Y'_{\beta rr}\beta r'^2 + Y'_{rrr}r'^3) \quad (3)$$
$$N_H = \tfrac{1}{2}\rho L^2 d U^2 (N'_\beta\beta + N'_r r' + N'_{\beta\beta\beta}\beta^3 + N'_{\beta\beta r}\beta^2 r' + N'_{\beta rr}\beta r'^2 + N'_{rrr}r'^3)$$

where :

$\beta = \tan^{-1}(v/u)$ and $r' = r(L/U)$

and heeling moment, Eq. 4 :

$$K_H = -z_H Y_H - B_{44}\dot{\phi} - C_{44}\phi \quad (4)$$

where :

$$z_H = OG - h$$
$$C_{44} = gm\overline{GM}$$
$$B_{44} = \frac{2}{\pi}\sqrt{gm\overline{GM}(I_{XX} + J_{XX})}$$

Acording to Kijima [8], Equation force and moment induced by propeller and rudder can be expressed by the Eq 5 :

$$X_P = (1-t_P)\rho K_T D_P^4 n^2 \qquad (5)$$
$$Y_P = 0;\ N_P = 0;\ K_P = 0$$

where:

$$K_T(J_P) = C_1 + C_2 J_P + C_3 J_P^{\ 2}$$
$$J_P = U\cos\beta(1-w_P)/(nD_P)$$

Force and moment on rudder area (X_R, Y_R, N_R and K_R)can be expressed by Eq. 6

$$X_R = -2(1-t_R)F_N \sin\delta$$
$$Y_R = -2(1+a_H)F_N \cos\delta \qquad (6)$$
$$N_R = -(x_R + a_H x_H)F_N \cos\delta$$
$$K_R = -z_R Y_R$$

where:

$$F_N = \tfrac{1}{2}\rho A_R f_\alpha U_R^2 \sin\alpha_R$$
$$f_\alpha = 6{,}13\Lambda/(2{,}25+\Lambda)$$
$$U_R = \sqrt{u_R^2 + v_R^2}$$
$$\alpha_R = \delta - \tan^{-1}\left(\frac{-v_R}{u_R}\right)$$
$$u_R = \varepsilon(1-w)u \times \sqrt{\mu\left\{1+\kappa\left(\sqrt{1+(8K_T/\pi J^2)}-1\right)\right\}^2 + (1-\eta)}$$
$$v_R = \gamma_R(v - r l_R)$$

2.2 PROGRAMMING AND SIMULATION

According to Maimun et al. [6], to analyze the flow of ship maneuverng motion, numerical simulation are based on the concept of time domain, where the path to be transersed by the ship was obtained from double integration of equations of motion of the ship.

$$Surge\ : X = m(\dot{u} - rv)$$
$$Accelerati\ \ on : \dot{u} = (X/m) + rv$$
$$Velocity\ : u = \int \dot{u}\, dt$$
$$Dis\ \tan ce : u = \int u\, dt$$

Then by the same method performed on the motion integration process sway, yaw and roll. Over the next simulation is developed and analyzed through computer simulation programs in MATLAB-Simulink software.

2.3 SHIP OPERATION SCENARIO

To determine the effect of rudder area on heeling manoeuvering of ship, the research will be on simulation scenario role in the operation of ships which are: 1) the vessel is operated with variation in rudder area (A_R = 1.632 m^2, A_R = 1.819 m^2 and A_R = 2.078 m^2). 2) The vessel is operated with variation of velocity (V = 2 knots, 6 knots and V = 10.5 knots). 3). Comparison between simulation results and the sea trial.

Tables 1, 2 and 3 respectively show the main parameters, the parameters of the propulsion and rudder, coefficient of KMP Sultan Murhum hydrodynamic derivatives. In support of maneuver ships, equipped with two conventional propulsion (FPP) and 2 conventional rudder is mounted behind the ship. A number of parameters in the prediction of propulsion using Holtrop Method [9][10],Rudder area parameters obtained of the field data, A_R = 2.078 m2 (original) and A_R = 1.632 m^2 (modified). Further hydrodynamic coefficients predicted by the derived regression equation developed by Yoshimura and Ning [11] and Yoshimura [12].

Table 1. Ship particulars

Parameter	Dimension	Unit
LOA	36.4	m
LBP	31.5	m
LWL	35.73	m
.B	8.7	m
H	2,.65	m
T	1.65	m
V	10.5	knot
Cb	0.63	
Δ	321.8	ton

Table 2. Propeller and rudder parameter

Parameters	Value
Dp, m	1.1
N	2
Z	2
A_E/A_O	0.40
RPS, n	8.578
w	0.219
t	0.142
Jp	0.499
K_T	0.230
A_R, m^2	2.078

Table 3. Calculated hydodynamic coefficients

Coeficient	Value	Coeficient	Value
		Y'β	0.4629
X'o	-0.00743	*Y'r-m'x*	0.0348
X'ββ	-0.1477	*Y'βββ*	1.2
X'βr-m'y	0.06604	*Y'ββr*	-0.5
X'rr	0.03	*Y'βrr*	0.34
X'βββ	1.183	*Y'rrr*	-0.04
N'β	0.1397	*1-tr*	
N'r	-0.05592	*ah*	0.8478
N'βββ	0.3	*ε*	1.0306
N'ββr	-0.33	*K*	0.3986
N'βrr	0.01	*l'R*	0.9042
N'rrr	0	*γr*	0.4884

3. RESULTS AND DISCUSSIONS

3.1.1 RUDDER AREA VARIATIONS.

For rudder area variations on turning circle maneuver, used three conditions, A_R = 1.623 m^2, A_R = 1.819 m^2 and A_R = 2.078 m^2 at a speed of 10.5 knot with a full draught 1.6 m. Simulation results can be seen in the fig 2:

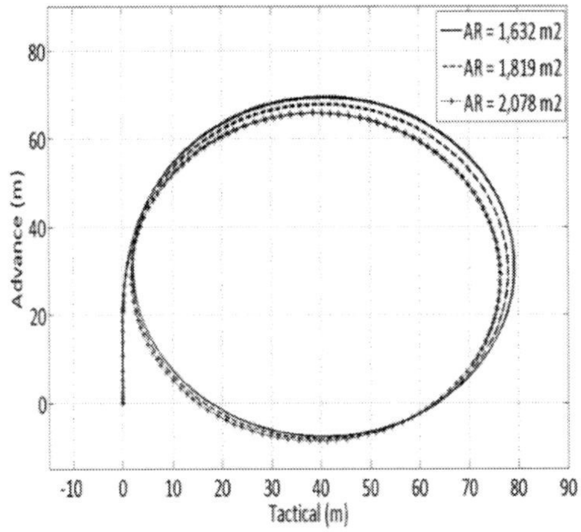

Fig. 2. Turning circle with rudder area variations

Based on Fig 2. it is known the value of tactical diameter $A_R = 1.632$ m^2 is 79.1 m, ie 2.51 L (31.5 m length of the ship). This value is larger than tactical diameter A_R= 1.819 m^2 which is equal to 77.9 or 2.47 L, and the tactical diameter A_R = 2.078 m^2 of 7.3 m or 2.42 L. While the advance value for A_R = 1.632 m^2 is 69.1 m or 2.2 L, greater than the advance for the A_R = 1.819 m^2 which amounted to 67.5 m or 2.1 L and A_R = 2.078 m^2 of 65.6 m or 2.1L. The larger rudder area, the steering force produced greater tactical so as the smaller diameter vessels. The detail can be seen in Table 4.

Table 4. The relationship between rudder area and ship maneuvering performance.

Parameter	IMO Criteria		Rudder Area (A_R), m^2		
			2.078	1.819	1.632
D_T, m	< 5 L	157.5	76.3	77.9	79.1
A_D, m	< 4.5 L	141.7	65.6	67.5	69.1

Based on Table 4, it is known that the tactical diameter and advance to variations in the use of rudder area still meets the criteria of IMO.

Table 5 shows the addition of tactical diameter of 3.5% for the use of A_R = 1.632 m^2 and 1.5% for the use of A_R = 1.819 m^2. While the addition of 5.1% advance for the use of A_R = 1.632 m^2 and 1.819 m^2 of use A_R = 52.4%.

Table 5.Percentage of tactical diameter and advancewith original rudder area (A_R) = 2.078 m2

Parameters	Rudder Area (A_R), m^2	
	1.632	1.819
D_T, m	3.5 %	1.5 %
A_D, m	5.1 %	2.3 %

Turning circle for the simulation results, also obtained information of drift angle, turning rate, and heel angle of ship that can be seen in Fig. 3, 4, and 5.

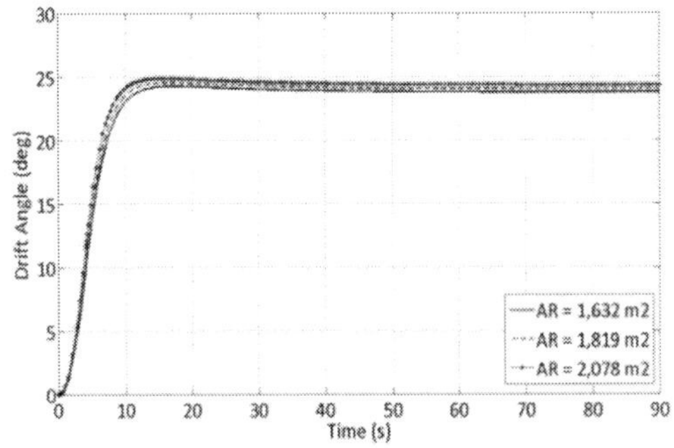

Fig. 3. Turning circle: drift angle

Based on Fig. 3, it can be seen that the drift angle A_R = 1.632 m^2 was 23.8^0, is smaller than the drift angle A_R = 1.819 m^2 which amounted to 24°, and the drift angle A_R = 2.078 m^2 of 24.4°. This is caused by the influence of v (Sway velocity) and u (surge velocity) are different for each of rudder area.

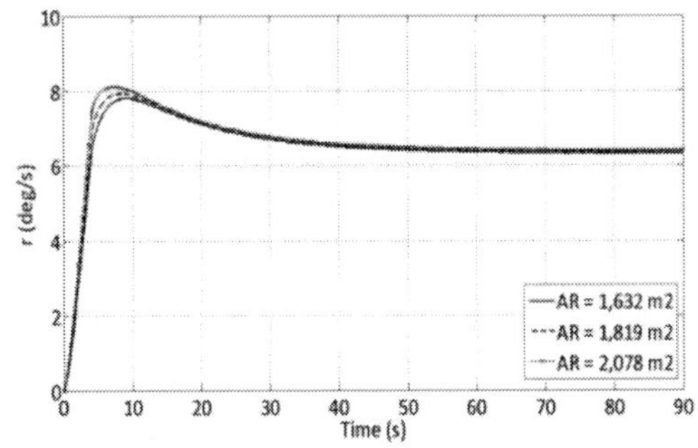

Fig 4. Turning circle: turning rate

Based on Fig. 4, it is known the value of the turning rate A_R = 1.632 m^2 is 6.4 deg/s, is greater than the turning rate of A_R = 1.819 m^2 which is equal to 6.1 deg/s, and the turning rate of A_R = 2.078 m^2 at 6 deg/s. This is caused by differences in the value of the drift angle, the moment induced by hull (N_H) and rudder (N_R) in the yaw moment.

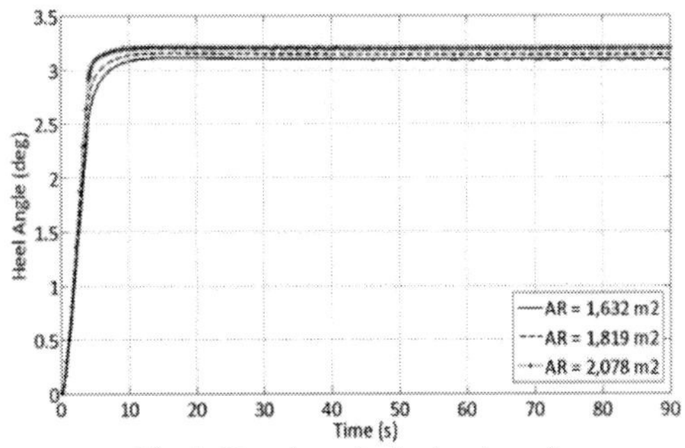

Fig 5. Turning circle: heel angle

Test results obtained A_R =1.632 m^2 has a heel angle is smaller than the A_R = 2.078 m^2 and A_R = 1.819 m^2, where the heel angle for A_R = 1.632 m^2 is 3,10 °, A_R = 1.819 m^2

is 3.15° and A_R = 2.078 m^2 at 3.21° as shown in fig.5. The simulation result of turning circle test shows that KMP Sultan Murhum have a good heeling of turning circle, this can be seen from the angle that still meets the criteria established by IMO (<10°). For the results can be seen in Table 6.

Table 6. The relation between rudder area to heeling maneuvering of ship

Parameters	IMO Criteria	1.632	1.819	2.078
Ø, rad	<10^0	0.7	1.1	1.49

For *zig-zag* maneuver can be seen in following fig. 6 :

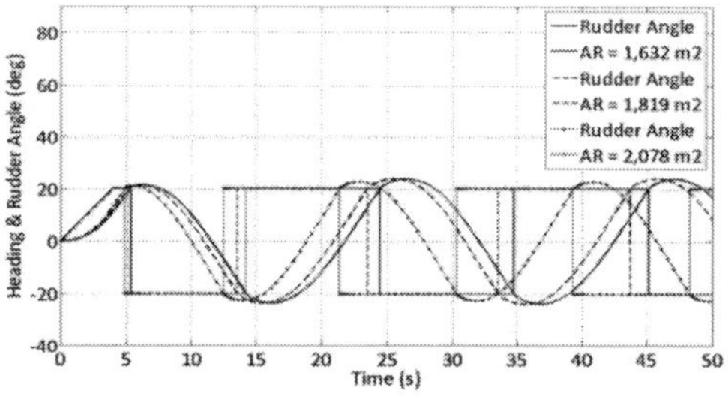

Fig 6. Zig-zag 20°/20°: rudder area variations

Fig. 6. shows the simulation results for the zig-zag manoeuver 20/20, where the horizontal and vertical axis shows the time and heading angle while maneuvering. For heading angle of A_R = 2.078 m^2 has smaller value of overshoot compared to the other rudder area.

For 1st *overshoot,* can be seen that for A_R = 1.632 m^2 the ship takes 6.3 seconds to turn right while for A_R = 1.819 m^2 and takes 5,9 seconds for the A_R = 2.078 m_2 takes time 5.5 seconds so it can turn right. Similarly, at the 2nd overshoot, for A_R = 2.078 m^2 takes 14 seconds faster to turn left than the other rudder area (A_R = 1.632 m^2 takes 16,5 seconds and A_R = 1.819 m^2 takes 15.7 seconds). For the results can be seen in Table 7.

Table 7. Relation between rudder area and ship manoeuver

Parameters	IMO Criteria		1,632	1,819	2,078
Zig-zag 20°, 1st Overshoot	< 25°	< 25°	1.5	1.4	1.2
2nd Overshoot	< 40°	< 40°	3.5	3.6	2.6

For zig-zag simulation result, obtained turning rate and heel angle can be seen in fig. 7 and 8.

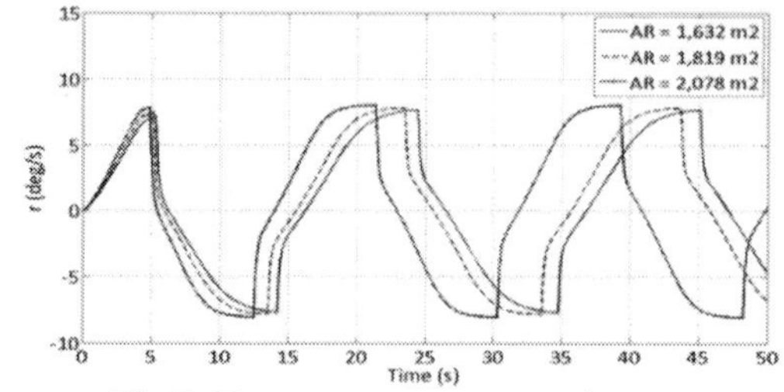

Fig 7. Zig-zag maneuver : turning rate

Based on fig.7 obtained turning rate when *zig-zag maneuver,* for turning rate A_R =1.632 m^2 greater than *turning rate* A_R =1.819 m^2 and *turning rate* A_R =2.078 m^2. This is caused by the differentiate of drift anlge result, moment induced by hull (N_H) and rudder(N_R) in *yaw moment.*

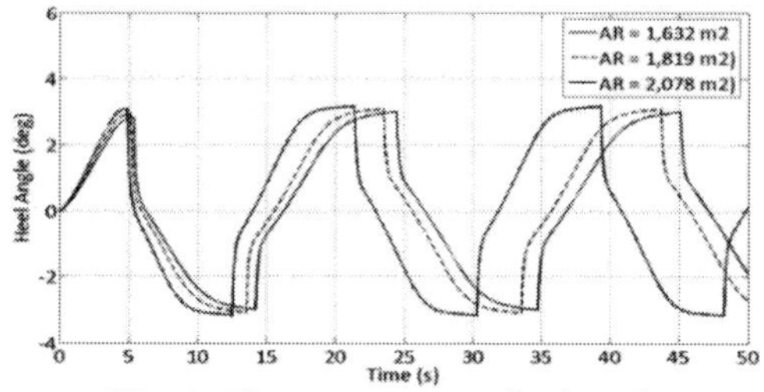

Fig. 8. Zig-zag manuver: heel angle

Base on fig.8 obtained heel when zig-zag maneuver, where heel angle of A_R = 1.632 m^2 smaller than the other rudder area.

3.1.2 SPEED VARIATONS.

Turning circle for maneuvering speed variation, we used three conditions, namely velocity (V) = 2 knots, V = 6 knots and V = 10.5 knots with draught 1.65 m. Simulation results can be seen in the fig.9:

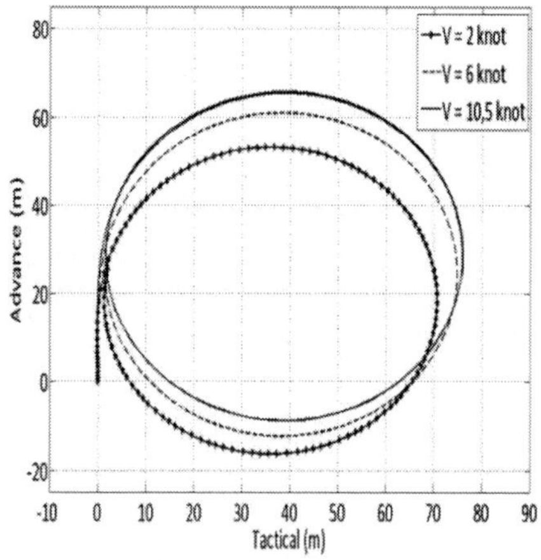

Fig. 9. Turning circle: speed variations

Based on Fig. 9, it is known the value of tactical diameter V = 2 knots is 70.7 m, which is 2.2L (about 31.5 m of ship length). This value is smaller than the value of tactical diameter V = 6 knots which is equal to 74.8 or

2.3L, and the tactical diameter of V = 10.5 knots is 76.3 m or 2.4 L. While the advance value for V = 2 knots is 53.8 m or 1.7L, less than the advance for V = 6 knots which amounted to 60.9 m or 1.9L and V = 10.5 knots at 65.6 m or 2.1L. The faster the ship, the forces induced by propeller (X_P which produced the more, so as the tactical diameter is greater. More detail can be seen in Table 8.

Table 8. Relation between speed and ship maneuver

Speed (V), knot	IMO Criteria		2	6	10,5
D_T, m	< 5 L	182	70.7	74.8	76.3
A_D, m	< 4.5 L	163.8	69.1	67.5	65.6

Reduction in the tactical diameter approximately 61.1% for V = 2 knots, 58.9% for V = 6 knots, and V = 10.5 knots at 58.1%. While the reduction in advance could reach approximately 67.1% for V = 2 knots, 62.8% for V= 6 knots, and 60% for V = 10.5 knots.

Turning circle for the simulation results, also obtained information inflection point of the boat (drift angle), angle the vessel per second (turning rate), and angle of heel that the ship can be seen in Figure 10,11 and 12.

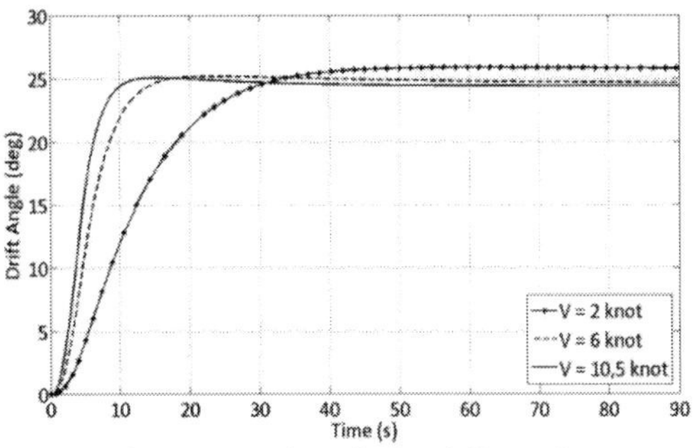

Fig 10. Turning circle: drift angle

Based on Fig. 10, it can be seen the drift angle V = 2 knots is 25.6 °, smaller than the drift angle V = 6 knots which is equal to 24.7 °, and the drift angle of V = 10.5 knots is 24.4 °. This is caused by the effect of surge velocity (u) and sway velocity (v) are different for each speed.

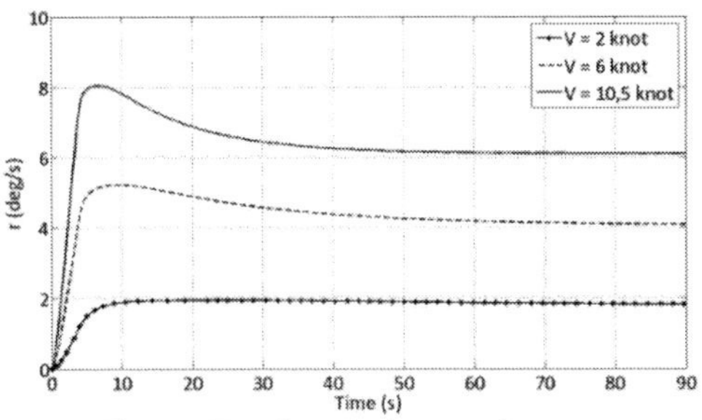

Fig 11. Turning circle : turning rate

Based on Fig.11, it is known the value of the turning rate of V = 2 knots is 1.8 deg/s, smaller than the turning rate V = 6 knots which is equal to 4.1 deg/s, and turning rate V = 10.5 knots at 6 deg/s. This is caused by differences in the value of the drift angle, the moment that induced by hull (N_H) and rudder (N_R) in the yaw moment.

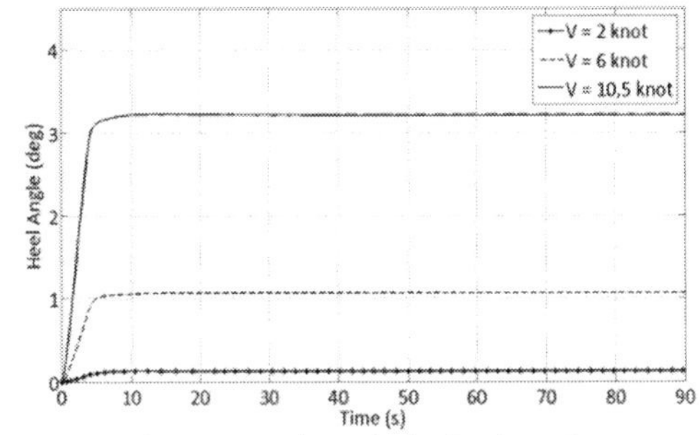

Fig 12. Turning circle: heel angle

The results shows the velocity V = 2 knots have a smaller heel angles than V = 6 knots and V = 10,5 knots, where the heel angle for V = 2 knots is 0,1°, V = 6 knots is 1,1° and V = 10,5 knots is 3.2°. The simulation results show KMP Sultan Murhum have a good heeling of turning circle, this can be seen from the angle that still meet the criteria established by IMO (<10°). For the results can be seen in Table 9.

Table 9. Relation between rudder area and heeling maneuvering

Speed (V), knot	IMO Criteria	2	6	10,5
Ø, rad	<10°	0.1	1.1	3.2

For zig-zag maneuver with speed variations can be seen in the following figure:

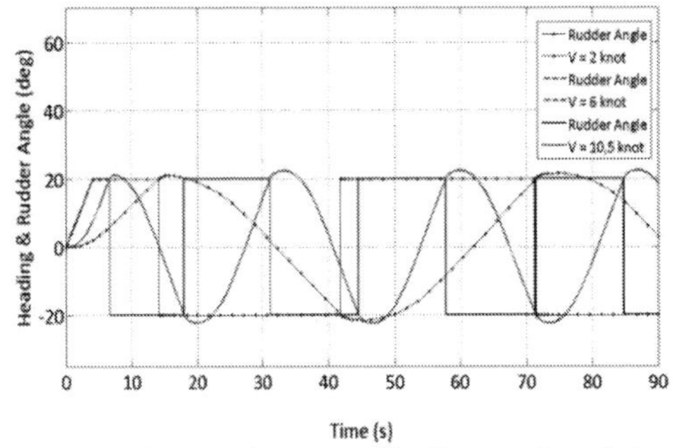

Fig 11. Zig-zag 20°/20° : speed variations

Figure 11 shows the simulation results for the zig-zag maneuver 20/20 where the horizontal and vertical axis shows the time and the current heading angle maneuver. For heading angle V = 2 knots has a value greater overshoot angle than the other speed .

For the 1st overshoot V = 2 knots, it can be seen that the ship takes 15.5 seconds to turn right while for V = 6 knots takes 7.4 seconds and to V = 10.5 takes 5.5 seconds to turn right. Similarly, at the 2nd overshoot, to ship with V = 10.5 knots takes 14 seconds to turn left faster than the speed of other vessels (V = 6 knots takes 19.8 seconds, and V = 2 knots takes 45.1 seconds). For the results can be seen in Table 10.

Table 10. Relation between speed variations and ship maneuver

Speed (V), knot	IMO Criteria		2	6	10,5
Zig-zag 20°, 1st Overshoot	< 25°	< 25°	1.0	1.2	1.2
2nd Overshoot	< 40°	< 40°	1.6	2.3	2.6

3.3 SIMULATION RESULTS AND SEA TRIAL

Ship simulation results using speed 10.5 knots, rudder area 2.078 m^2 and draught 1.65 m obtained tactical diameter value by 76.3 or 2.4 L, while based on sea trial testing the boat on turning circles and zig-zag maneuvers, the tactical diameter of KMP Sultan Murhum is 46.5 m or 1.5 L smaller than the simulation results. For the results can be seen in Table 11.

Table 11. Comparison between the Sea Trial with simulation results

Turning Circle Test		Simulation	Sea Trial
Rps, n	Rpm	2100	2100
A_R	m^2	2.078	2.078
DT	M	76.3	46.5
BHP	HP	1100	1100
δ	o	20	20

After the simulation with a variety of rudder area and ship speed can be known for A_R = 2.078 m^2, which can be simulated minimum speed is 2 knots, where the tactical diameter of 70.7 m. To get the tactical diameter of 46.5 simulation is not possible, so that the expected speed of the ship when the test is done by using a speed below 2 knots, or not at full speed The simulation of turning circle with speed V = 2 knots can be seen in figure below :

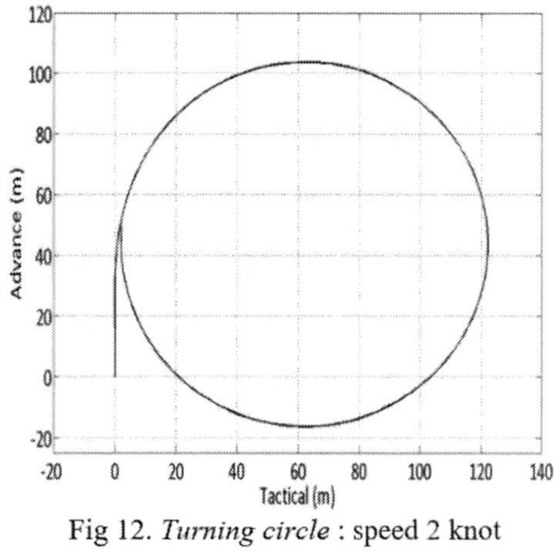

Fig 12. *Turning circle* : speed 2 knot

4. CONCLUSIONS

Based on the analysis, the following conclusions have been made.

1. The larger the rudder area the greater the steering force produced. This is in line with smaller tactical diameter and advance.
2. Ferry of *KMP SULTAN MURHUM* has a good heeling of turning circle as the heel angle meets the IMO criteria.

5. ACKNOWLEDGEMENTS

The authors wish to thank Ministry of Culture and Education of the Republic of Indonesia for funding this research.

6. REFERENCES

1. Noor, D.C.H.B.M., 2009 '*Manoeuvring Prediction of Offshore Supply Vessel'*. Faculty of Mechanical Engineering, UTM, Malaysia.
2. Muhammad, A.H. and Paroka,D. 2007 , *'Pengembangan Program Simulasi Manuver Domain Waktu Model Planning Hull (Kapal Patroli',,* Seminar Nasional Teori Dan Aplikasi Teknologi Kelautan (SENTA), Indonesia.
3. Zaojian, Z.2006, Lecture Notes on Ship Manoeuvring and Seakeeping, Jiao Tong, Shanghai University.
4. Viviani, M, et al. 2003, '*Identification of hydrodynamic coefficient from standard manoeuvres for a serie of twin-screw ships'*. Genova, Italy.
5. IMO. '*Standards for Ship Manoeuvrability'*.2002 Report of the Maritime Safety Committee on its Seventy-Sixth Session-Annex 6 (Resolution MSC.137(76)*)*. London.
6. Maimun, A., Muhammad, A.H. and Salem, A.. 2004, '*Development of A Simulation Program for Pusher-Barge Manoeuvring'*, 9th JSPS Marine Transportation Engineering Seminar. Hiroshima, Japan.
7. Ogawa, A. and Kansai, H., '*On the Mathematical Model of Manoeuvring Motion of Ship'*, International Shipbuilding Progress. Vol. 25, No. 292, pp. 306-319. 1987
8. Kijima, K,. Yasuaki, N,. '*On the Prediction Method for Ship Manoeuvring Characteristics'*, Proceeding of Marsim 2003, Japan.
9. Holtrop, J. and Mennen G.G.J.1982, An Approximate Power Predition Method, International Shipbuilding Progress, Vol 29.
10. Holtrop, J. 1984, A Statistical Re-analysis of Resistance and Propulsion Data, International Shipbuilding Progress, Vol 31.

11. Yoshimura, Y., and Ning Ma.,2003 '*Manoeuvring Prediction of Fishing Vessels'*, Proceeding of Marsim , Japan.
12. Yoshimura. Y., 2010 '*Principle of the effect of roll motion on ship manoeuvring dynamics'*.(In Japanese). Proceedings of the Conference of Japan Society of Naval Architects and Ocean. Japan.

7. AUTHORS' BIOGRAPHY

N, Yuniarsih, M.T. She was born in Kendari, Indonesia in 1983. She graduated with a B.Sc. degree in Naval Architecture, Hasanuddin University, Indonesia in 2007. He obtained his Masters from the same university in 2012 in the field of ship manoeuvering.

Dr. A. Haris Muhamamad. He was born in Bandung, Indonesia in 1969. He graduated with a B.Sc. degree in Naval Architecture, Hasanuddin University, Indonesia in 1993. He obtained his M.Eng and Ph.D. in 2001 and 2007 from Technology Sepuluh Nopember of Institute (ITS) and Malaysia University of Technology respectively. He joined the Faculty of Engineering, Hasanuddin University as a lecturer staff in 2000. He is currently the Deputy Head of Department of Naval Architecture, Fac. Of Eng.,Hasanuddin University.

Dr. M. A. Djabbar holds the current position of head of Masters study program in Naval Engineering, Hasanuddin University, Indonesia. His research interests include high speed boat, mainly hydrofoil and ocean environment, particularly ocean wave mechanics. He obtained B.Sc.(Naval Arch.), Hasanuddin University, in 1976,.M.Eng (Naval arch.) Hiroshima U., Japan in 1981 and Doctor (Ocean Eng.) in 2002, Bandung Institute of Technology, Indonesia respectively. He joined the Faculty of Engineering, Hasanuddin University as a Lecturer in 1976 and appointed as Professor in 2007. He is a member of international committee, MARTEC conf. .

MARINE SAFETY IN INDONESIAN WATERS AND SOME INSIGHTS GAINED FROM THE EUROPEAN EXPERIENCE

R. Birmingham, Newcastle University, UK
K. Utama, ITS Surabaya, Indonesia
Moch. Zaky, Research & Development, PT. Biro Klasifikasi Indonesia - Persero
H. Enshaei, Newcastle University, UK

SUMMARY

Indonesia is one of the largest archipelago countries in the world in which two thirds of its territories consist of sea areas with nearly 18,000 islands and a coast line 81,000 km long. It has a population of two hundred million people. Consequently sea transportation plays a very important role in this nation's development, complementing the land and air modes of transportation. Many types of sea transportation operate in Indonesian waters carrying people and cargo between the islands, and harvesting the natural marine resources.

As with other modes of transport the marine infrastructure in Indonesia is continually developing. This is the result of growing economic activity and with this growth an increase in socio-cultural community activities. In addition changes in the national regulation of transportation has led to an increase in all transport activity, but for the marine sector a consequence of this increase is that their has also been an increase in the incidence of marine casualties. Bad weather is inevitably more likely to result in a casualty if the vessels involved have inadequate seaworthiness and are overloaded. This unfortunately is too often the case in Indonesia. This unsatisfactory situation can be traced to two factors in the development of Indonesia's maritime infrastructure.

The first of these is the condition of the fleet. Many vessels are second-hand purchases from other countries, and maintained to an unsatisfactory standard. The second problem is the weak supervision of shipping safety standards, commonly resulting in situations such as overloading or the carrying of unreported hazardous goods. While poor weather may often be the initiating factor in a casualty, good forecasts of weather conditions from the Meteorological, Climatology and Geophysics Agency (MCGA) mean that this should rarely be an issue if the regulation of ship seaworthiness and operating practices raised the standard of safety.

1. INDONESIAN FLAG-SHIPS

Based on the ships registered at the Ministry of Sea Transportation of Indonesia, the number of vessels in Indonesia in April 2010 was 52,890. Of these vessels 46,076 ships were less than 500 gross tonnes while 6,814 vessels of more than 500 gross tonnes, as illustrated in Figure 1.

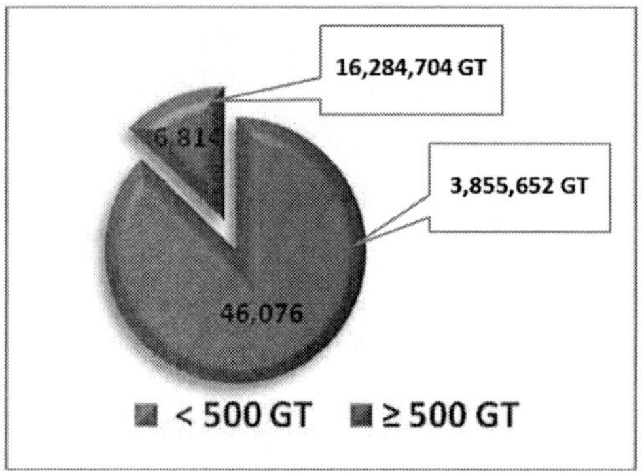

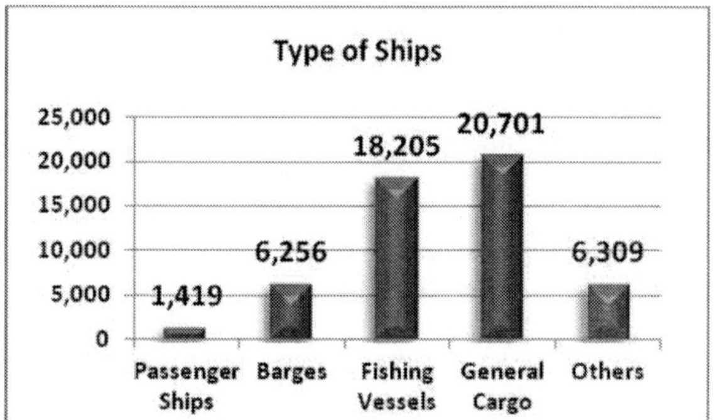

Figure 1. The number of Indonesian flag-ships up to April 2010

Four of the most common types of vessel operating in Indonesian waters are briefly described below, with an outline of the problems associated with such vessels:

- Traditional boats are generally made of wood and do not have regulatory standards of seaworthiness imposed. This type of ships is most frequently involved in accidents caused by overloading, combined with designs that have not been subject of proper scrutiny of their stability.

Figure 2. Traditional Boat

- Ro-Ro ferrys operating in Indonesia commonly do not have watertight car decks because of the many openings on all sides of the ship. As a result if the intact and damage stability is calculated it is found that the IMO criteria are not met. This can be due to inadequate reserve buoyancy being provided by the car deck when the ship is heeled or flooded. Incidents that have occurred on Ro-Ro ferries have been caused by overloading, fires or explosion resulting from vehicles on board carrying hazardous cargos, and water ingress to the car deck caused.

Figure 3. Ro-Ro Ferry

- Landing Craft (LCT) are modified in order to perform as commercial ferries carrying both passengers and vehicles. These ships operate between islands where port infrastructure is minimal or non existent. Commonly the ships have little reserve buoyancy with no watertight compartment above the main deck. In addition the domestic rules for the freeboard calculation do not include a bow height correction so the value of the freeboard minimum is significantly different from the freeboard calculation for the International Load Line Convention of 1966.

Figure 4. Landing Craft

Cargo ships, many of which are second hand vessels purchased from Japan. In many cases on transfer to Indonesian ownership the freeboard deck will be changed from the tween deck to the upper deck in order to increase the draft and so carry more cargo. With this change the vessel no longer meets International Load Line regulations. Even without such a change these vessels are old, and have no subdivision of the cargo space, so that damage causing flooding of the hold compartment inevitably leads to grounding at the very least, and more likely to sinking.

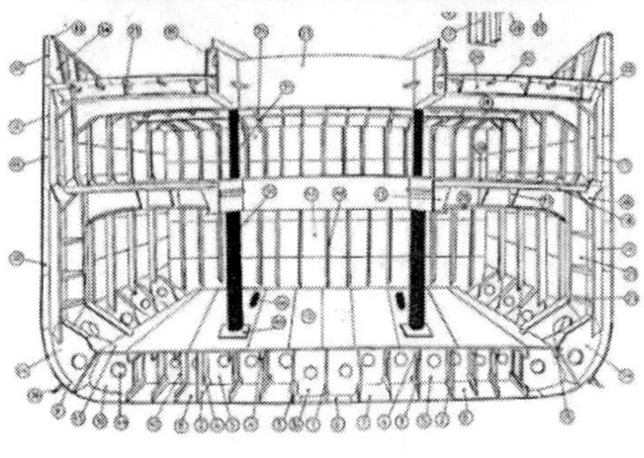

Figure 5. Typical of cargo ship with tween deck

2. MARINE CASUALTIES IN INDONESIAN WATERS

If an event suffered by a ship 'could threaten the safety of the ship and/or human life' then, according to the Shipping Act of 2008 (Act No 17, Article 245), the Admiralty Court undertakes a full investigation. Examples of such casualties are shown in Figures 6, 7 and 8.

Figure 6. Sinking and Capsizing: both the cases of the passenger ship MV. Windu Karsa and the MT. Josephine were caused by water ingress into car deck.

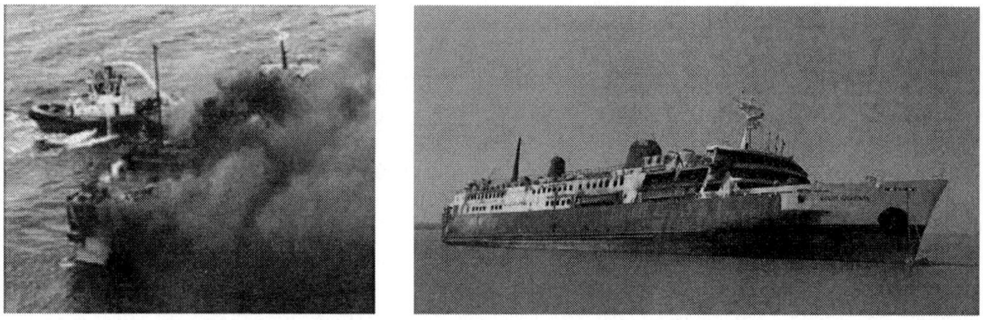

Figure 7. Fire and Explosion: the cause of the fires on the passenger ship MV. Levina 1 and MV. Mandiri Nusanatara were due to dangerous goods being transported by vehicles on the car deck

Figure 8. Collision: the collision between the container ship MV. Unichart and MV. Mandiri Nusanatara in Madura strait

Based on the data from Directorate General of Sea Transportation during the period 2007-2011 there have been 27 marine casualties in Indonesian waters. The break down of the reason for these casualties is that sinking caused 37%, fire or explosion 41%, and collision 22%. An alternative breakdown of these marine casualties in Indonesia is that during the period 2007 - 2011 59% were due to technical factors and 41% were due to human factors (NTSC, 2011). This analysis is shown graphically in Figure 9.

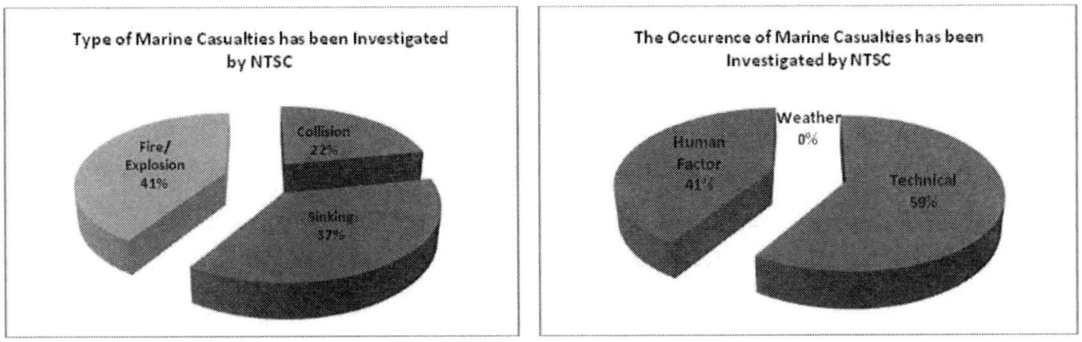

Figure 9. Marine Casualties in Indonesian waters period 2007-2011 (NTSC, 2011)

While the data taken from the Director General of Sea Transportation provides a useful analysis (shown in Figure 9) it is evident that this source provides an incomplete picture. An earlier investigation by the National Transport Safety Committee (NTSC or KNKT), and the Indonesian Coastguard (KPLP), found that during the period 2003-2008 there had been 724 marine casualties in Indonesia waters. Analysis of this data, shown in Figure 10, indicate that most of the marine casualties in Indonesia waters occur on traditional boats with the most common type of accident being sinking. It should be noted however that this categorization does not shed light on the underlying cause of the sinking event.

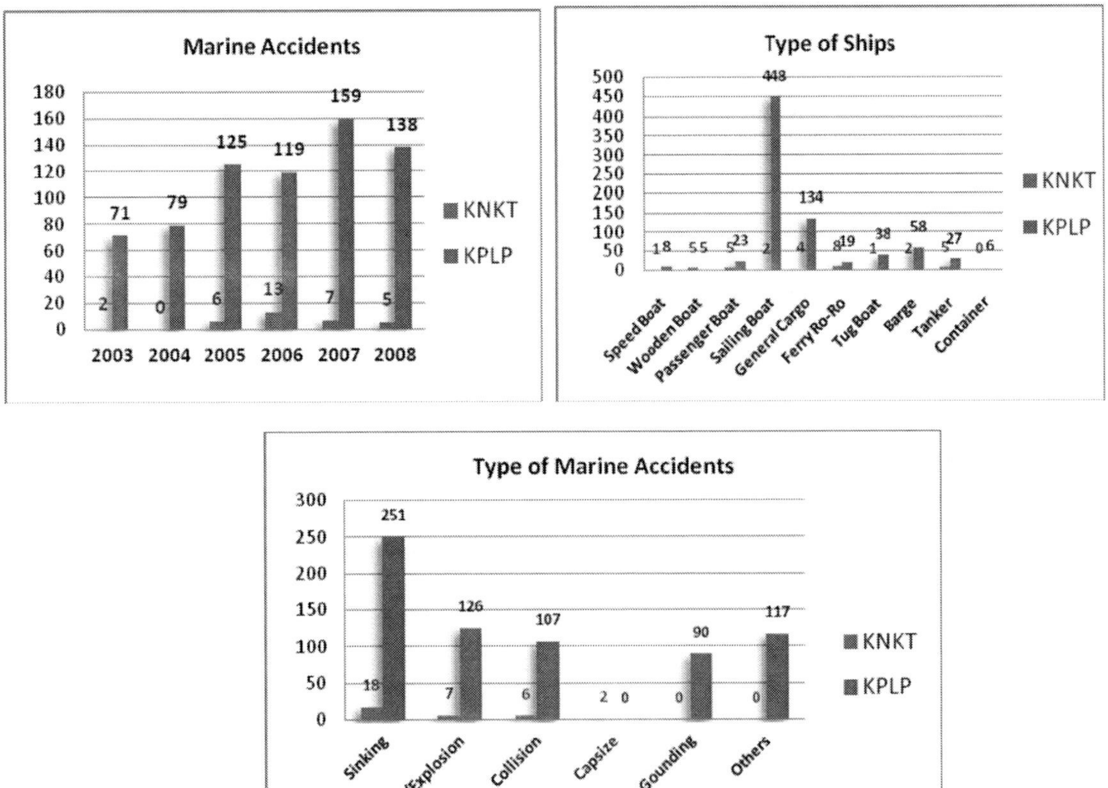

Figure 10. Marine Casualties in Indonesian waters period 2003-2008 (NTSC, 2008)

3. DEVELOPMENT OF RISK BASED APPROACH TO SAFETY IN EUROPE

The conventional approach to safety in the design of ships (and the operational practices associated with them) has been to retrospectively ensure adequate measures are in place. This can be seen both in the design of individual vessels, where an evaluation of the safety of the vessel would be undertaken near the end of the design process, and in the implementation of regulations, many of which have been introduced as a result of a safety failure. The most significant example of this was the initiation of the SOLAS regime following the loss of the Titanic. However in recent decades in the marine industry there has been a recognition that safety must be undertaken in parallel with other design activity, and regulations implemented in anticipation of the likelihood of casualties (Birmingham et al, 2000, 2004). In the 1990s this change in approach was referred to as 'Design for safety', however in recognition of the reality behind such a process the term now used is 'Risk based design'.

In the marine industries this approach became a major area for research in Europe in the 1990s. In other industries (such as the nuclear industry) where safety, or more precisely the lack of safety, impacts directly on the wider community the issue of risk has a longer history. Figure 11 shows how the risk based approach to safety has grown, and from this figure it can be seen how specific disasters, many of them in Europe, were the spur to the development of risk based design.

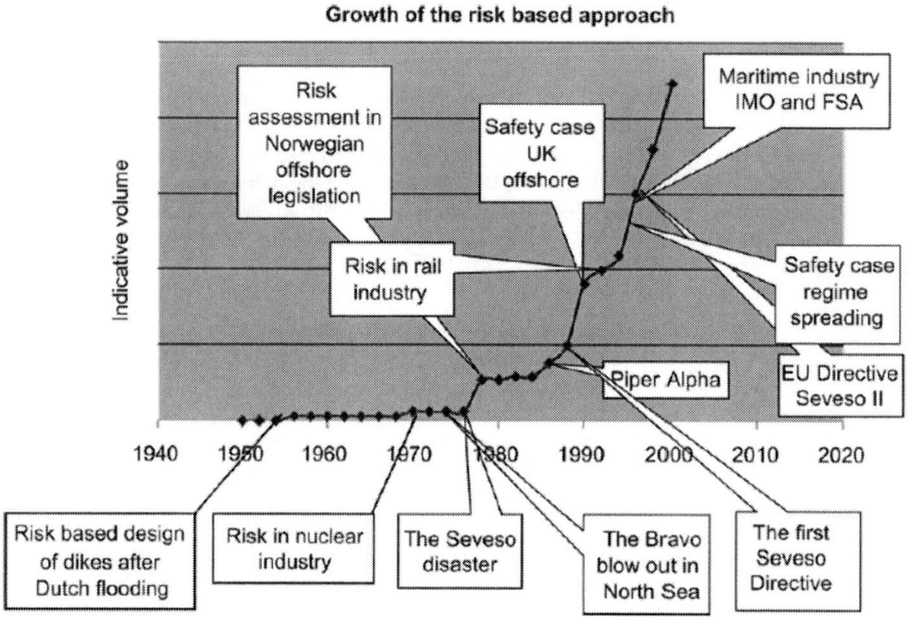

Figure 11 Historical development of risk based approach (Pasman et al, 2009).

One approach to embed safety considerations into the heart of the design process was to undertake formal safety assessment as a parallel activity, running in tandem with established design practice. With this approach it was proposed that a safety model of the design was generated that could be used to evaluate the impact of all design decisions, and so inform the designer of the relative safety of alternative proposals (Birmingham et al, 2000, 2004).

Formal safety assessment has five stages: hazard identification, risk analysis, risk control options, cost-benefit analysis, and recommendations for decision making (IMO, 2007). To undertake this effectively can be resource intensive, meaning that it is expensive in terms of time. The risk analysis stage in particular requires a detailed evaluation of the propagation of failures (often done with fault trees) and the likelihood of consequences. Undertaking this analysis in a qualitative way can be challenging for large complex systems, however it is valuable as identifying the failure pathways can indicate where a system lacks robustness and is vulnerable to specific malfunctions. A more complete picture is obtained where a quantitative analysis is undertaken, with failure and consequence probabilities being established, and impacts evaluated. The data required to undertake such an analysis is often not available, and may either be impossible to obtain or only with considerable effort. In the past this only seemed justified in industries such as the nuclear and aviation industry, however society is becoming less tolerant of risk, as demonstrated by the willingness of governments and international agencies to fund major research projects in the area of marine safety. In the last two decades many projects have been undertaken in Europe to develop tools for the analysis, evaluation and management of risk, so enabling decisions concerning safety to properly take account of risk, and this done at a cost that is acceptable. These research projects are also accumulating the data that is necessary to quantitatively evaluate risk in the maritime world.

One project funded by the European Commission has analysed casualties in order to establish the risk associated with various casualty scenarios. This project, called FIRE EXIT, has been reported on by Vanem and Skjong (Vanem and Skjong, 2004a,b), and all of the risk models shown below will refer to the risk models developed in the FIRE EXIT project. The high level risk analysis that they report on allows an estimation of the potential benefits that are possible if the risk can be reduced. The analysis was limited to ships engaged in international trade, and only ships greater than 4000 Gross Tonnage (GT) were considered. Cruise ships and Ro–Ro passenger ferries were included in the study, using casualty statistics for the years 1990–2006.

Risk models developed in the FIRE EXIT project are reproduced in Figures 12 to 16. These models are deemed most relevant, as the focus of the FIRE EXIT project was on the emergency evacuation from passenger ships (Vanem and Ellis, 2010).

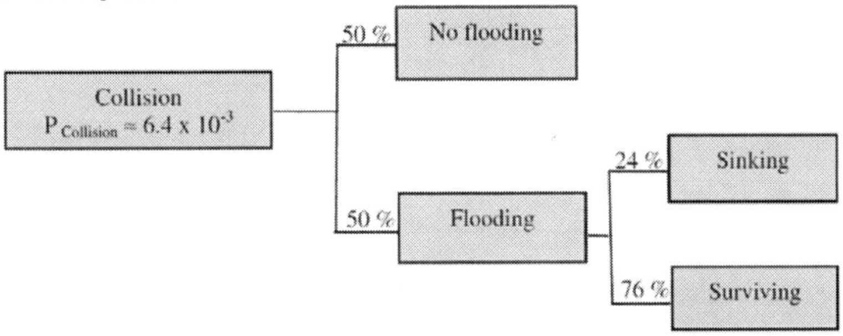

Figure 12: Risk model for collision of passenger ships.

Available time for evacuation (min)	<5	5-10	10-15	15-30	30-60	60-90	>90
Probability	0.23	0.24	0.16	0.17	0.08	0.06	0.06
Expected fatality rate (%)	96	88	80	63	40	20	7

Table 13: Probability estimates for time to sink and expected fatality rates in the case of collision.

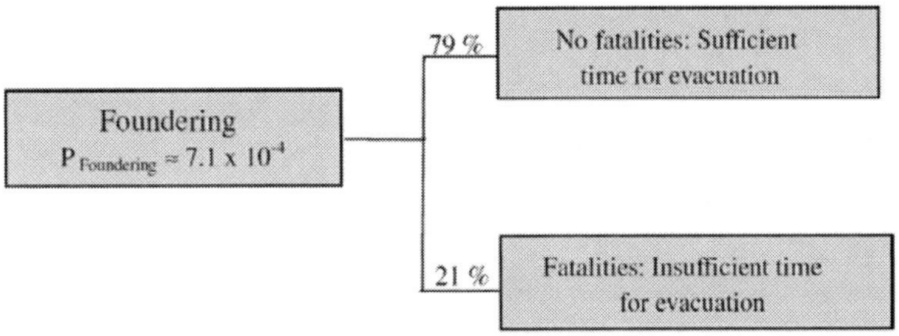

Figure 14: Risk model for foundering of passenger ships. The frequency of fatal foundering accidents involving passenger ships is 1.5 x 10 $^{-4}$ per shipyear.

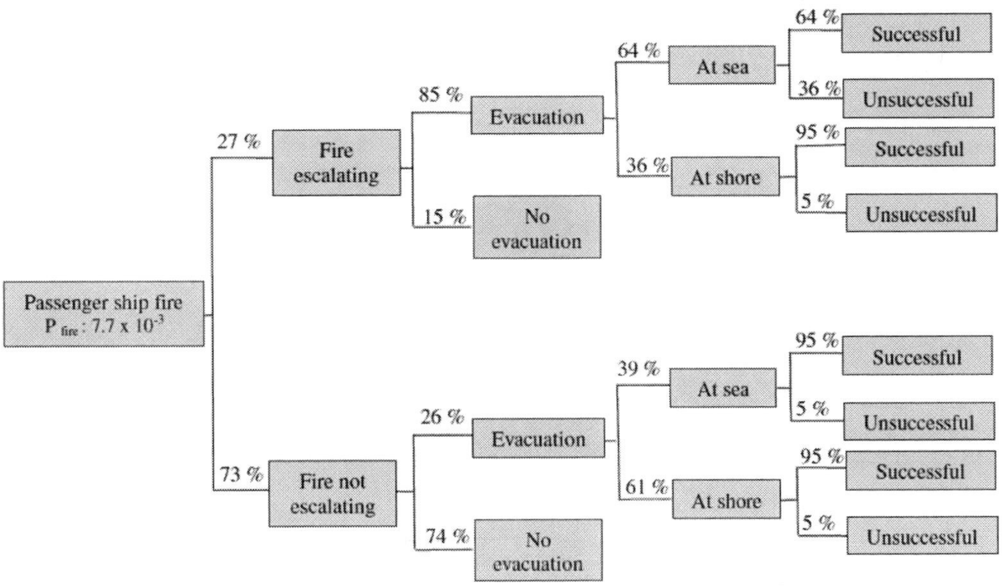

Figure 15: Risk model for fire and explosion of passenger ships

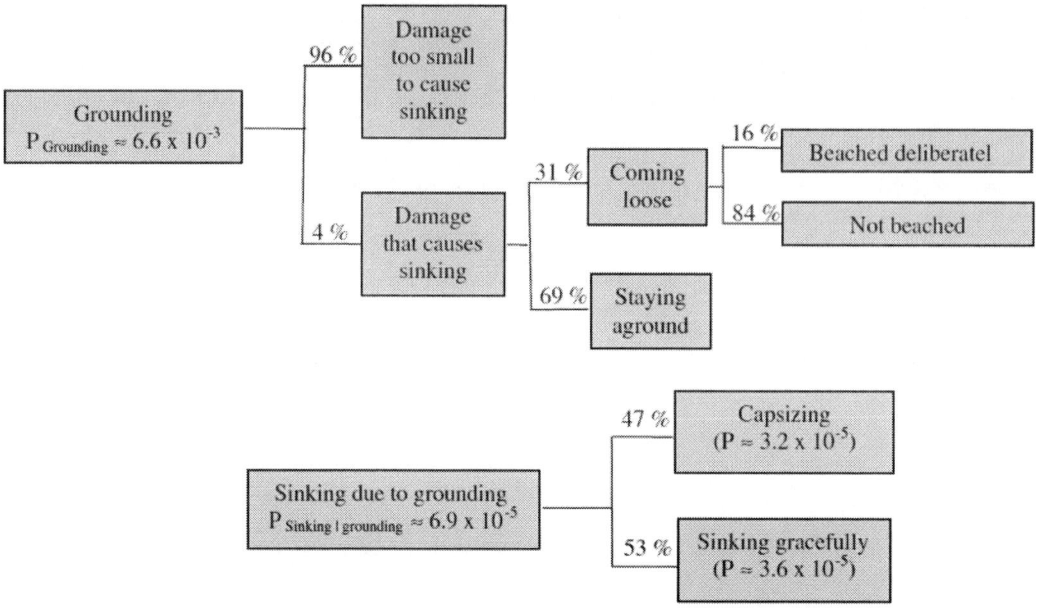

Figure 16: Risk model for grounding of passenger ships

4. PROGRESS IN INDONESIA TOWARD A MARINE SAFETY FRAMEWORK

The Indonesian Classification Bureau, known as BKI (Biro Klasifikasi Indonesia), was founded 1964 with its main function being that of a Classification Society: surveying vessels being operated in Indonesian waters and classifying them according to their condition. BKI is dedicated to enhancing safety at sea, protecting the marine environment from ship pollution, and developing, promoting and assisting in the implementation of relevant International standards. Some classification activities are based on the ship's hull construction, machinery and other systems, with the aim of providing an assessment of the seaworthiness of the vessel to sail. BKI is authorized by the Government to carry out statuary survey and certification activities on behalf of the Indonesian Government, such as Load Line, ISM and ISPS Codes. BKI also surveys and issues certificates for the national load line rule (KM 3 year 2005) and International Regulation (ICLL 1966).

BKI maintains a register of ships which, in the light of the discussion above, can provide valuable information on the Indonesian fleet. This data set could be used as a starting point for the development of a risk model for Indonesian shipping, paralleling the results on the international fleet reported on by Vanem and Skjong. Figures 17 and 18 provide a simple breakdown of the fleet in terms of type, tonnage, and age.

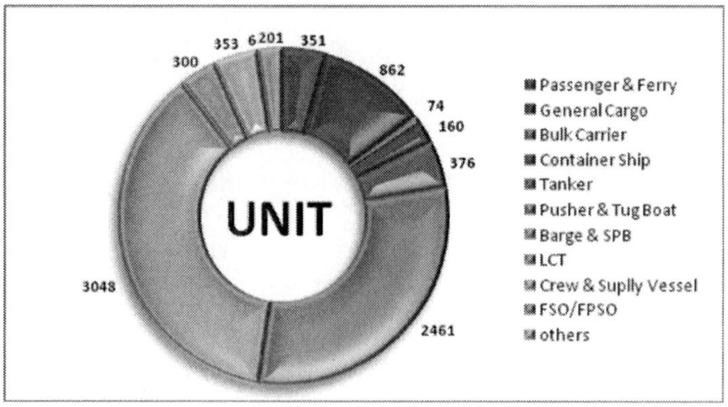

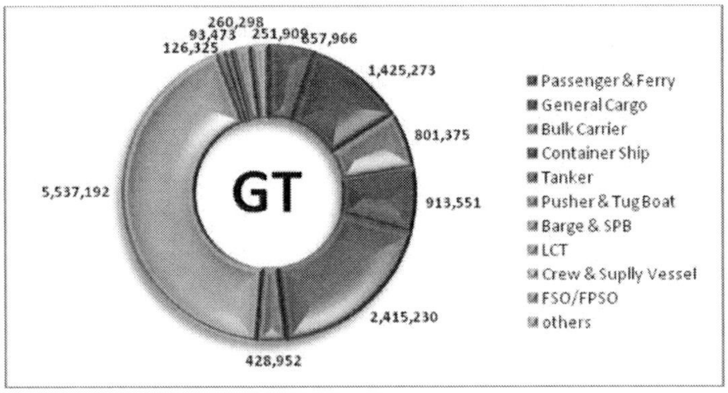

Figure 11. Ships registered by BKI. In July 2012 the register had 8192 vessels with a total of 12,911,545 gross tonnes.

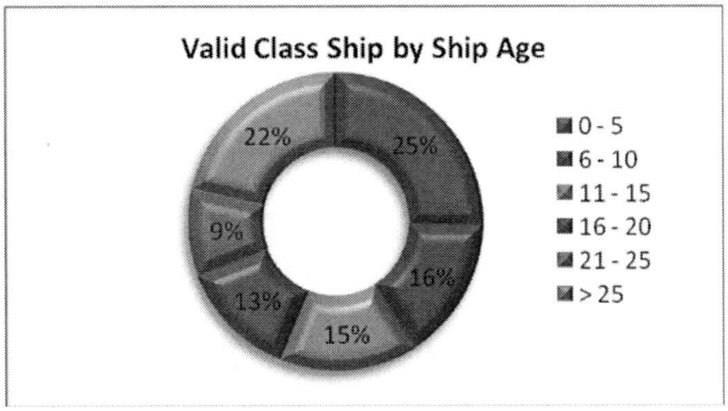

Figure 12. Ship age registered by BKI

As discussed at the beginnng of this paper the Indonesian fleet comprises a large number of small traditional vessels, many built of wood, and these are disproportionately represented in the casualty statistics. Between 1997 and 2006 a research programme was undertaken as a collaborative project between Indonesia and Japan, the JSPS-DGHE Programme. The purpose of this research program was to reduce shipping accidents by developing a domestic Indonesian rule for the safe operation of ships in Indonesian seas. The results of this study included the application of seakeeping theory, based on the Japanese domestic rules, which could be applied in Indonesia. By calculating the distribution of wave height and period for the territorial waters of Indonesia, and combining this with motion characteristics collected for ships operating in different areas of Indonesia, the proposed rules were intended to reduce vessel casualties.

In 2009 the Ministry of Transportation introduced the Non Convention Vessel Standard, known as NCVS (Ministry of Transportation No. KM 65 year 2009) for vessels registered under the Indonesian flag that are not provided for in international conventions. This Standard states that it will apply to the following vessel types:

- All merchant vessels on domestic voyages
- Merchant vessels less than 500 gross tonnes on international voyages
- Vessels with non mechanical propulsion (barge, pontoon, and sailing vessel)
- Wooden vessels with sail and auxiliary propulsion engine (KLM) and wooden vessels with propulsion engine
- Fishing vessels
- Pleasure craft
- Vessels of novel design
- State vessels which are used for commercial purpose
- All existing vessels which have been modified for another purpose.

From this list it can be seen that this Standard is specifically designed to address the large Indonesian fleet of small vessels that are not catered for by most such codes. It is notable that it specifically mentions wooden vessels, fishing vessels and sailing vessels, which numerically are a large proportion of the Indonesian fleet.

5. CONCLUSION

As an archipelago nation, with a population of 200 million people, Indonesia is, and always will be, dependant on its maritime infrastructure to ensure a dynamic economy and healthy social and cultural regional interactions. The vessels that provide the necessary transportation links, and that commercially harvest the natural resources of the nation, are varied in size and type. Ongoing development will enhance the standard of the fleet, but at this time many of the operating vessels are second hand, modified for the local conditions, and operate with inadequate maintenance. The combination of low operating standards and insufficient inspection has resulted in casualties that could be prevented.

In this paper the Indonesian experience is contrasted with the attitude to safety that has developed in Europe, where the aversion to risk is increasing year on year, with the result that more resources are being invested in understanding and implementing risk based approaches to the design of vessels and their operation.

It is noted that such procedures are resource intensive, particularly in terms of the person hours required to gather the necessary data and to conduct the meaningful analysis that will improve safety. However it is also recognized that organizations such as the Indonesian Classification Bureau, BKI, are already gathering relevant data that could be used to establish risk models for the Indonesian fleet. Such models indicate the probabilities of loss as a result of various scenarios. Such analysis should guide the development of future rules and the implementation of new operational practices. This will ensure that efforts to change the design of vessels, and the habits of the seafaring community of Indonesia, will be focused on the areas that will have the greatest impact on safety.

REFERENCES

- Birmingham, R.W.; Sen, P.; Cain, C.; Cripps, R.M. (2000), 'Development and implementation of a design for safety procedure for search and rescue craft.' *Journal of Engineering Design*, Vol 11, No 1, March 2000, p. 55-78.
- Birmingham R.W., McGregor J., Delautre S., Astrugue J. A., (2004), 'Risk Evaluation at the Preliminary Design Stage of High Speed Craft', *Journal of Ship Production*, Volume 20, No. 3, Society of Naval Archtiects and Marine Engineers, New Jersey, August 2004.
- IMO (2007), "Consolidated text of the Guidelines for Formal Safety Assessment (FSA) for use in the IMO rule-making process", MSC/Circ.1023-MEPC/Circ.392.
- Indonesian Bureau Classification (2012), BKI Ship Register 2012, BKI Jakarta
- JSPS-DGHE (2006), Final Report for Collection of Wave Data and Safety of Ships Operating
- Ministry of Law and Human Rights of Indonesia (2008), Shipping Act No. 17 year 2008, Shipping Regulation, Jakarta
- Ministry of Sea Transportation of Indonesia (1999), Presidential Decree No. 105 year 1999 concerning National Transportation Safety Committee, Jakarta
- Ministry of Sea Transportation of Indonesia (2010), Indonesian Ship Register Year 2010, Jakarta
- Ministry of Transportation of Indonesia (2009), KM 65 year 2009, Non Convention Vessel Standard Indonesian Flagged, 1st Edition, Jakarta
- National Transportation Safety Committee (2008), Sea Transportation Analysis Report 2003-2008, NTSC Final Report, Jakarta
- National Transportation Safety Committee (2011), Analysis of Accident Data and Investigation of Sea Transportation Year 2007-2011, Media Release of NTSC, Jakarta
- Pasman, H.J.; Jung, S.; Prem, k.; Rogers, W.J.; Yang, X. (2009), "Is risk analysis a useful tool for improving process safety?", Journal of loss prevention in the process industries, Vol 22, p. 769-777.
- Vanem, E.; Ellis, J. (2010), "Evaluating the cost-effectiveness of a monitoring system for improved evacuation from passenger ships", Journal of safety science, Vol 48, p. 788-802.
- Vanem, E.; Skjong, R. (2004a), "Fire and evacuation risk assessment for passenger ships", In: Proc. 10th International Fire Science and Engineering Conference (Interflam) 2004, Edinburgh, Scotland, July 5–7, Vol 1, p. 365–374. ISBN: 0 9541216-4-3.
- Vanem, E.; Skjong, R. (2004b), "Collision and grounding of passenger ships – risk assessment and emergency evacuations", In: Proc. 3rd International Conference on Collision and Grounding of Ships, ICCGS 2004, Izu, Japan, October 25–27.

BIOPOLYMER OF MONEPTERUS ALBUS SLIME ON CATAMARAN MODEL WALL FOR TURBULENT DRAG REDUCTION

M .A. Talahatu, Sunaryo, Warjito and Yanuar, Department of Mechanical Engineering, University of Indonesia

SUMMARY

The surface viscous liquid is one of fish's resistance reduction elements. The fish surface is covered with viscous liquid, which is considered to reduce the resistance. As compared with a solvent, a certain polymer solution of several ppm to several hundred ppm offers substantially reduced turbulent frictional resistance. The relationship between the Tom's effect and reduction of viscosity resistance is described. The purpose of this research is to study the drag reduction in resistance of catamaran model by slime monepterus albus solution in wall. The catamaran model with hull L = 1.8 m, B = 0.32m and T = 0.12 m pulled by electric motor which the motor speed can be varied. The ship model resistance is measured by load cell anemometer. The difference of friction coefficient is expressed as a reduction in friction drag. The results shows the drag reduction about 4 -7%.

NOMENCLATURE

υ	Kinematic viscosity (N s m^{-2})
ρ	Density of water (kg m^{-3})
C_T	Total coeffisien resistance (-)
L	Length of ship model (m)
Lpp	Length between prependicular (m)
B	Breadth (m)
T	Draft (m)
V	Velocity (m/s)
C_F	Friction coeffisien resistance (-)
C_W	Wave coeffisien resistance (-)
$(1+k)$	Form factor (-)
τ	Wave interference factor (-)
β	Viscous interference factor (-)
S	Wetted surface area (m^2)
Fr	Froudes number (-)
Re	Reynolds number (-)
g	Acceleration of grafity (m/s^2)
DR	Ratio drag reduction (%)

1. INTRODUCTION

Many researchers have been focus attention on drag reduction phenomena in turbulent flow with additives solution. Biopolymers such as high molecular weight polysaccharides produced by living organisms can provide effective drag reduction. Polysaccharides of several fresh water and marine algae, fish slimes, seawater slime and other fresh water biological growths have been found to be good drag reducers [1]. Interestingly, as mentioned later these biological additives are also a source of fouling growth which can substantially reduce the drag reduction effectiveness brought about by other drag reduction technologies. The external layer of slime on fish was investigated as a polymer. Virtually all fishes including eel are covered with slime secretion that is involved in many aspects of their biology, ranging from disease resistance and rearing of young to shelter and locomotion [2]. In locomotion, the layer of slime is assumed to reduce friction drag. Dissolved in water, these external secretions have been shown to reduce friction drag in turbulent pipe flow [3]. These results have been attributed to the "Toms effect" in which dilute solutions of high-molecular-weight polymers flowing in pipes yield friction-drag coefficients less than values predicted for Newtonian fluids. Most authors agree that fish slime reduces drag. No one, however, has estimated the degree to which slime can decrease the friction drag experienced by swimming fish. The drag-reducing ability of fish slime in solution has been determined by a turbulent flow rheometer [4]. In addition, Ling and Ling [5] investigated the velocity field in the laminar sub-layer of slime injected into water flowing in a tube. In all these experiments, slime samples were removed from fish and analyzed for drag reduction as a function of slime concentration. In reality, slime is secreted by the integument and diffuses into water at some unknown rate.

Most authors agree that fish mucus reduces drag. Hoyt [6] tested the slime of a sea fish and of a hagfish and found small drag reductions of 14.5% and 12.8%, respectively. Recently, Yanuar *et al* [7] investigated experimentally the drag reduction effects due to the introduction of slime solution into a mono-hull ship at the basin. The study established a relationship between the drag reduction effectiveness and the slime solution position along the hull. Yanuar *et al* [7] found that the slime was effective for drag reduction. The slime appear to have very strong effects on dynamics of turbulence for drag reductions greater than about 8%.

In particular, the interaction between the slime and the boundary layer on twin hull such catamaran has not been studied extensively. The present study is to investigate experimentally the drag characteristics using slime on a displacement catamaran. A catamaran model with dimensions of L = 1800 mm, B = 320 mm, and T = 120 mm is used. The influence of the varied location slime placed is also investigated.

2. EXPERIMENTAL DESCRIPTIONS

A series of model tests were conducted in a basin. The basin had a length of 50 m and width of 10 m, and the water depth was maintained at a constant depth of 2 m. The experiments were conducted for a Froude number of up to 0.6. The model was connected to the load cell transducer at a point located amidships and vertically above the base line, allowing the model to move freely in the vertical plane. Total resistance was measured for each run over the test range of Froude numbers. In the resistance tests, the ship model was pulled by a wire rope and the total longitudinal force acting on the model was measured for various speeds. During the measuring run the ship model was free to heave and pitch. Fig.1 shows the experimental setup in the basin.

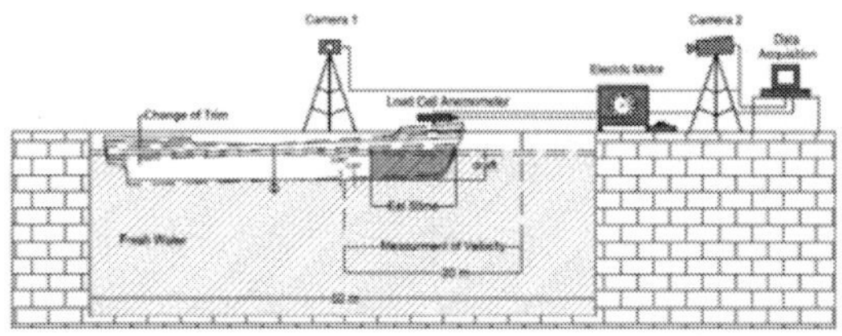

Fig 1: Experimental set-up

This set-up consisted of the catamaran models, electric motor, data interface, camera, load cell transducer and compressor. The comparison of the total drag between catamaran models with slime and without slime placed was analyzed. The model test was conducted in order to have the total resistance values of the catamaran model (R_T) at various velocity conditions (V). During the model test experiments, the catamaran model was pulled by an electric motor that was designed so that motor rotation could be used to pull the catamaran model with a constant speed. Pull force was measured by using a load cell transducer. The load cell was affixed to the ship model and connected to the rope pullers. The load cell was mounted on the amidships of the model. Towing rope was connected to an electric motor whose speed could be set and adjusted. The load cell gauge was connected to a data interface to obtain the pull force when the model was pulled.

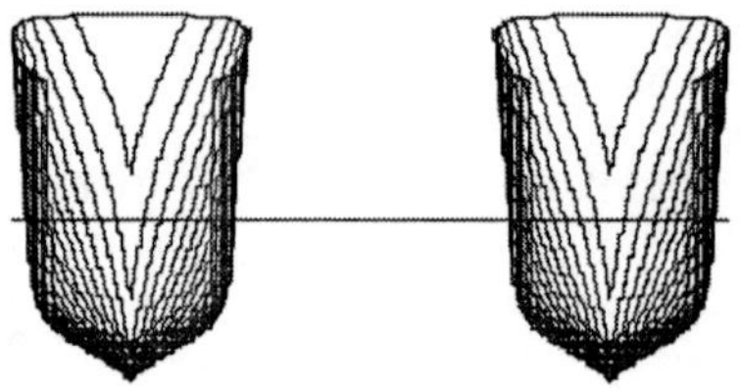

Fig 2: Lines and body plan of catamaran

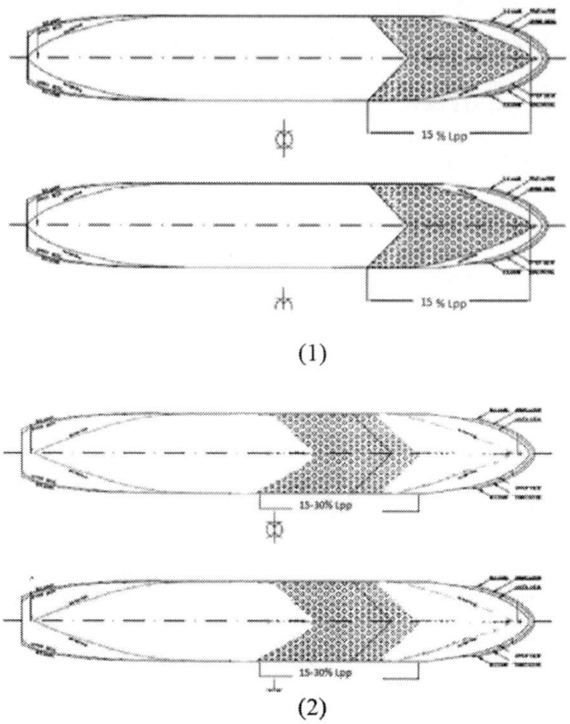

Fig 3: Positioning of slime placed

Fig.2 shows the lines plan and body plan for test model. Fig. 3 shows positions varied from position 1, position 2, and position 3. These positions were investigated and compared regarding the influence slime placed. There are three types of slime placed configuration. Position 1 is placed around the 0-15% Lwl from the bow. Position 2 is placed around the 15%-30% Lwl from the bow and Position 3 is placed around the 30%-45% Lwl from the bow.

3. EXPERIMENTAL ANALYSES

This paper constitutes a resistance analysis of symmetrical displacement catamaran hull forms with various lateral separations (clearance) on the effects of slime added application. The paper by Insel and Molland [13] summarizes a calm-water-resistance investigation into high-speed semi-displacement catamarans, with symmetrical hull forms based on experimental work carried out at the University of Southampton. Two interference effects contributing to the total resistance effect were established; these are viscous interference, caused by asymmetric flow around the demihulls which affects the boundary layer formation, and the wave interference, due to the interaction of the wave systems produced by each demihull. They proposed that the total resistance of a catamaran could be expressed by the equation:

$$(C_T)_{CAT} = (1 + \varphi k)\,\sigma C_F + \tau C_W \quad (1)$$

The factor ø has been introduced to take account of the pressure-field change around the demihulls and σ takes account of the velocity augmentation between the hulls and would be calculated from an integration of the local frictional resistance over the wetted surface, while (1+k) is the form factor for the demihull in isolation. For practical purposes, ø and σ can be combined into a viscous interference factor β where $(1+øk)\ \sigma = (1+\beta k)$. Hence

$$(C_T)_{CAT} = (1 + \beta k)\, C_F + \tau C_W \quad (2)$$

For a demihull, τ (wave interference factor) and β (viscous interference factor) =1.

Where C_T is coefficient of total resistance, C_W is coefficient of wave resistance, C_f is coefficient of friction resistance, and $(1+\beta k)$ is the form factor.

The form factor "$1+\beta k$" is determined experimentally at very low speed or low Froude numbers ($Fr < 0.2$), where Cr must become negligible. The ITTC [14] and Bertram [15] recommended applying Prohaska's method, described as:

$$C_T = (1 + k)\, C_F + a Fr^n \quad (3)$$

At low speed, $Fr < 0.2$, is assumed to be a function of Fr^4, the straight line plot of C_T/C_F versus Fr^4/C_F will intersect the ordinate ($Fr = 0$) at $(1+k)$, enabling the form factor to be determined. Flat plate friction resistance (C_F) is estimated with a reasonable precision using ITTC-57.

From the experimental towing test results, for the models, the total resistance coefficient C_T has been calculated as:

$$C_T = \frac{R_T}{0.5 \rho S V^2} \quad (4)$$

Where ρ is water density and S is the wetted area of the ship hull.
The Froudes number and Reynolds number are defined as

$$Fr = \frac{V}{\sqrt{gL}} \quad (5)$$

$$\mathrm{Re} = \frac{VL}{\nu} \quad (6)$$

Where V is the speed of the ship, L is the length of the ship, g is acceleration of gravity, and ν is the kinematic viscosity of water.

Drag reduction is obtained by:

$$DR(\%) = \left| \frac{C_T - C_{TO}}{C_{TO}} \right| \times 100\% \quad (8)$$

where C_{TO} is the total coefficient resistance of monohull and C_T is the total coefficient resistance of catamaran model. The catamaran model is varried by 4 types (1 types without slime placed and 3 types with slime placed).

4. RESULTS AND DISCUSSION

A comparison with the bubbles and without the bubbles applied to the catamaran model was made to evaluate the resistance characteristics as shown in Figs.4-6.

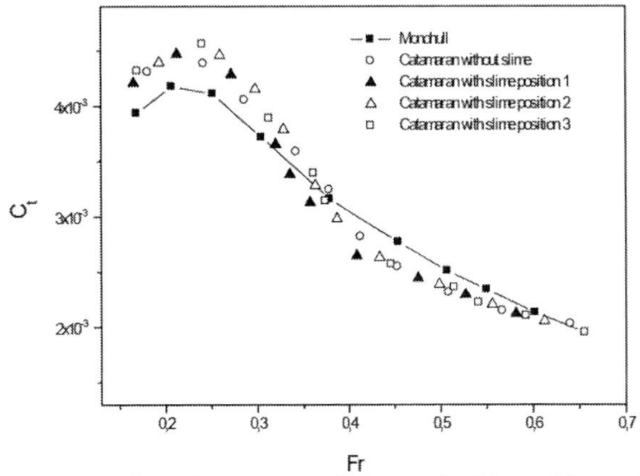

Fig.4 Total resistance coefficient of ship with and without slime placed

Fig.3 shows the relationship between the total resistance coefficient and Froude number for catamaran model with slime, catamaran without slime and monohull model. The catamaran model with slime placed will be varied for 3 variations of positioning. It appears that the catamaran model (with slime or without slime) has a value of C_T, which is relatively higher at low speeds. When the Froude number increases further, Fr > 0.35, in a certain range of values of C_T smaller than the monohull model. The effects can be seen in the slime at high Froude numbers where the turbulent flow drag coefficient of resistance is smaller. On the other hand, it was found that the value of C_T depends on the position of the slime placed. It appears that the ship model with slime at position 1, has a smallest value of C_T. It can be concluded that the slime position has an effect on the value of the resistance coefficient. For all models, the total coefficient reistance is increasing up to Fr about 0.25. The peak of C_T is describe that in Fr = 0.25 the flow occure maximum of fluctuation so value of C_T is maximum.

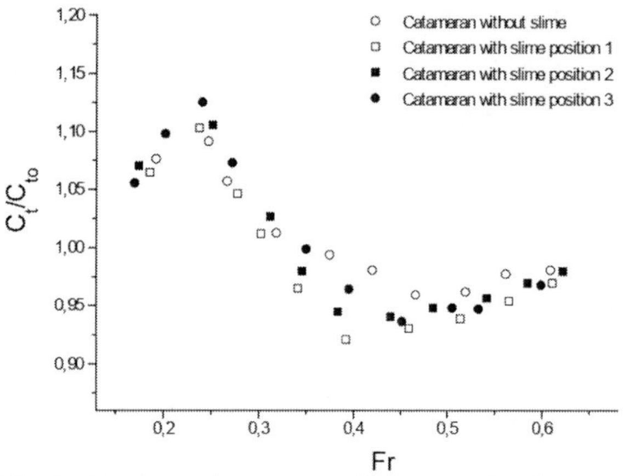

Fig.5 Total resistance coefficient ratio between catamaran model and monohull model.

Fig.5 shows the total resistance coefficient ratio as a function of Froude number. The value of ratio indicate that drag reduction occured. If the ratio value is lower than 1, it is indicated that resistance coefficient of the catamarn model is lower than monohull model. The best configuration is position 1 because have smallest resistance ratio.

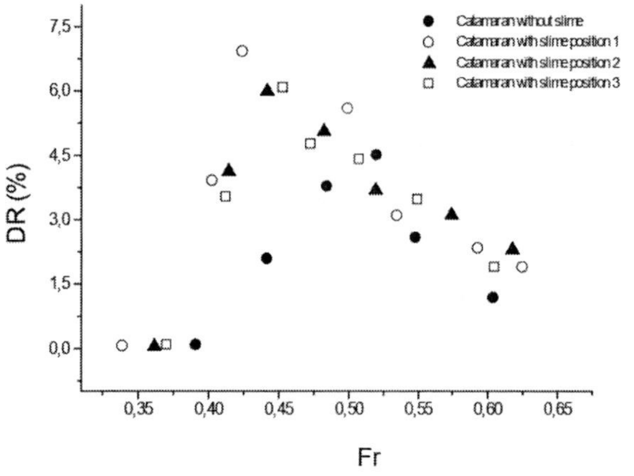

Fig 6: Relationship between drag reduction and Froude number

Fig. 6 shows the drag reduction that occurred. It is clear that drag reduction for position 1 is greater than position 1, position 2 and without slime. The drag reduction starts at a Froude number of about 0.35. As the Froude number increases, drag reduction also increases for spesific range. The start drag reduction occured is different for all variation. Catamaran with slime position 1 has start occured at Fr = 0,35. For position 2, 3 and without slime are follow respectively. The maximum drag reduction for this research is 7%.

5. CONCLUSIONS

Considering the experimental model test results to evaluate the slime drag reduction on a catamaran model, the following conclusions can be stated; slime application to ships is found to have a positive influence on catamaran resistance. The applications' of slime can reduce the catamaran resistance significantly for a high speed range. The test results show that an effective drag reduction of up to 7% at Fr = 0.45 can be achieved. The location of the slime position are significant for achieving the maximum drag reduction. Slime position at the stem area of bottom is the best location for effective drag reduction. It is expected that more work will be carried out in order to gain further understanding on this topic, in order to apply the method to full-scale ships

6. ACKNOWLEDGEMENTS

This work is supported by the Directorate for Research and Community Service, University of Indonesia (RUUI Research 2012), Jakarta, Indonesia.

7. REFERENCES

[1] HOYT, J. W. 1975. *Hydrodynamic drag reduction due to fish slimes*. Pp. 653-673 in T. V. T. Wu, C. J. Brokaw, and C. Brennon, Eds., *Swimming and flying in nature*. Plenum Press, New York.

[2] JAKOWSKA,S. 1963. *Mucus secretion in fish-a note*. Ann. N. Y. Acad. Sci. 160: 458-462.

[3] HOYT, J. W. 1974. Recent progress in polymer drag reduction. Colloq. Int. Cent. Natl. Rech. Sci. 223: 193-215.

[4] ROSEN, M. W., AND N. E. CORNFORD. 1971. Fluid friction of fish slimes. Nature 234: 49-51.

[5] LING, S. C., AND T. Y. LING. 1974. Anamolous drag-reducing phenomenon at a water/fish-mucus or polymer interface. J. Fluid Mech. 65: 499-512.

[6] Hoyt. J . W.. and Fabula. A. G. "The effect of additives on fluid friction."Naval Ordnance rest Station. China Lake, Calif., 1964. (Navweps Report 8636.)

[7] Yanuar, Talahatu, Sunaryo, Warjito. 2011. Drag Reduction in Ship Model by Biopolymer of Slime Monepterus Albus Solution in Wall. Int'l Jurnal Mekanikal Univ. Teknologi Malaysia. No. 32, 96 – 107

8. AUTHORS' BIOGRAPHY

M.A. Talahatu received his B.S. degree in Naval Architecture from Hasanuddin University, Indonesia, in 1982. He then received his Master of Eng. Mechanical Engineering from University of Indonesia, in 2003. Now, he is continuing Doctor program in Mechanical Engineering at University of Indonesia in Jakarta, Indonesia. His research project is about drag reduction in vessel with slime biopolymer.

Yanuar received his B.S. degree in Naval Architecture from University of Indonesia, Indonesia, in 1986. He then received his Master of Eng. and Doctor of Eng. in Mechanical Engineering from University of Tokyo Metropolitan, Japan, in 1995 and 1998, respectively. Professor at Mechanical Engineering at University of Indonesia in Jakarta, Indonesia. His research project is about drag reduction in internal flow or eksternal flow.

Sunaryo received his B.S. degree in Naval Architecture from University of Indonesia, Indonesia, in 1982. He then received his Master of Phil. And Ph.D in Naval Architecture from University of Strathclyde, Glasgow, U.K. in 1989 and 1993. Now, He is a Head of Program in Naval Architecture and Marine Engineering in University of Indonesia.. His research project is about Ship Production Management.

Warjito received his B.S. degree in Mechanical Engineering from University of Indonesia, Indonesia, in 1988. He then received his Master of Eng. and Doctor of Eng. in Mechanical Engineering from University of Hokkaido, Japan, in 1999 and 2002, respectively. His research project is about bubble generator and application.

STUDY ON THE POSSIBILTY OF ESTABLISHING SHIPBUILDING CLUSTER IN LAMPUNG PROVINCE SUMATRA INDONESIA

Sunaryo, University of Indonesia, Indonesia

SUMMARY

As the result of the implementation of cabotage principle in 2005 by the government, the number of national fleet has increased dramatically around 5,000 units in the last five years due to import of vessels from abroad, which mostly are relatively old and need renewal, maintenance and repair. This situation creates potential market for new building as well as repair yards in Indonesia. In response to this growing market the research is aimed to identify the possibility of establishing a shipbuilding cluster in Lampung Province at the southern part of Sumatra Island as complementary to the existing shipbuilding centers in Batam, Jakarta, East Kalimantan and Surabaya to cater the demand, and to make the most of industries around it. The study is started by gathering primary and secondary information regarding the location, market, resources and infra structure, and supporting industries, simulating the information, draw conclusions, and propose a conceptual design.

1. INTRODUCTION

As the result of the implementation of Cabotage principle in 2005 by the government of Indonesia through the Presidential Instruction no.5, 2005 on the Enabling of National Shipping Industry [1] the number of national cargo shipping fleet has increased significantly around 60% from 6,041 ships in March 2005 to 10,884 ships in September 2011[2]. Even though the increase is mainly due to import of used ships and more than half of them are reasonably old, ranging from 15 to 20 years of age [3]. This means that in 5 to 10 years time the fleet needs renewal, on the other hand in the mean time the existing fleet also needs routine maintenances and repairs as required by classification and statutory rules and regulations.

This booming market is a golden opportunity for the national shipbuilding and other related industries, but unfortunately the existing shipbuilding industry has not yet ready to anticipate to the promising market in term of its capacity, quality, delivery time, and cost of work. In response to this the local government of the Province of Lampung in the southern part of Sumatra Island based on its geographical location and condition, industrial and economical development in the neighbouring regions, the growing of transportation routes that connect Java and Sumatra, and the growth of the educational and training institutions in the province is planning to establish an integrated shipbuilding cluster as complementary to shipbuilding centres that have already existed such as in Batam, Jakarta, Surabaya and East Kalimantan.

The study was conducted to investigate the visibility of realising this project plan, which covered market opportunity, geographical aspect, industrial aspect, and commercial aspect.

2. SHIPBUILDING CLUSTER

Industrial cluster is defined as a group of inter-related industries that drive wealth creation in a region, primarily through export of goods and services. It represents the entire value chain of a broadly defined industry from suppliers to end products, including supporting services and specialized infrastructure. Cluster industries are geographically concentrated and inter-connected by the flow of goods and services [4].

In the case of shipbuilding cluster there are 4 industries that inter-related with shipbuilding industry as the core, they are: pulling industry, supplying industry, supporting industry, and supporting services [5], which can be illustrated as follows:

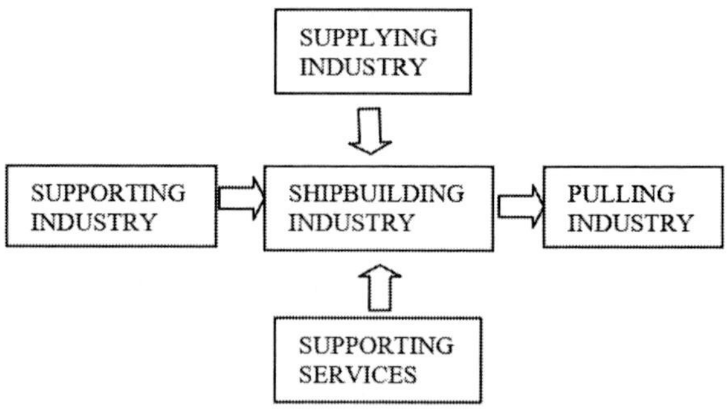

Figure 1: Shipbuilding Industry Cluster

Shipbuilding industry is the core of the cluster that will create wealth in the region with all its prospective opportunities, shipping industry is the pulling industry that provide market for the shipbuilding industry, supporting industry is the industry that produces main materials and components for shipbuilding that need further fabrication or work by the shipyards such as steel plates and sections and other hull materials, fabrication industry, and consumables such as water, gas and electricity etc., supplying industry is those that supply all kind of machineries and equipment for the ship that only need to be installed or applied by the shipyards including paint, chains, zinc anodes etc., and supporting services are institutions that provide services both directly and indirectly to the shipbuilding industry such as classification societies, design and surveying firms, educational institutions, financial institutions etc. All the inter-related industries should commit themselves to foster the development of the shipbuilding cluster.

3. PROJECTED MARKET OPPORTUNITY

Prior to the investigation to the possibility of establishing an integrated shipbuilding cluster in the Province of Lampung the size of projected market opportunity was firstly identified in order to decide the optimum size of the proposed shipbuilding cluster which would include the identification of the projected market demand of new-building and repair works, the existing condition and future development of shipbuilding industry in Indonesia, and the market size that might be captured.

3.1 PROJECTED MARKET DEMAND

In conjunction with the growth of Indonesian economy and due to implementation of Cabotage principle the number of national shipping fleet is also growing. From March 2005 to September 2011 the fleet has grown to nearly 11,000 units. 70% of this fleet are considerably old, ranging from 15 to 20 years of age and even 25 years, therefore in the next 5 to 10 years the ships need to be replaced by the new ones. If it is assumed that the demand is increasing by 5% each year the demand for new fleet is around 550 units every year and if 25% of it is new building there are 137 units of new building in demand every year. Beside the demand for replacement the fleet also needs maintenance and repair in order to fulfil the requirements of classification and statutory rules and regulations. If it is assumed that 50% of the fleet are maintained and repaired locally there are almost 6,000 units of ship queuing in the national repair market. According to INSA (*Indonesia National Ship-owners Association*) it is predicted that there are around 17 millions GT of ships need to be repaired and 700 to 1,000 units or equal to 1 million GT of new building [6].

3.2 EXISTING CONDITION OF SHIPBUILDING INDUSTRY IN INDONESIA

Refer to the record of the Ministry of Industry there are around 260 shipyards in existence in Indonesia with the capacity of 600,000 GT for new building and 9.5 millions GT for repairs [7]. The capacity utility of the national shipbuilding industry is 95% for repair and 50% for new building [8]. The low utility for new building works is mainly due to the low competitiveness of the industry in term of building cost, quality, and delivery time.

The main reasons of the un-competitiveness among other things are: out of date production process facilities and equipment, under qualified human resources, high imported ship components cost, inappropriate production system, and lack of government financial support.

Based on the availability of the shipbuilding and ship repair market, financial support, and ease of obtaining the ship components Indonesian shipbuilding industries are concentrated in three main regions i.e. Batam Island, Java Island, and East Kalimantan. Some new shipyards are sprouting in these three regions mainly for building tugs and barges for carrying coal, and offshore support vessels, and large ship repair yards for tankers and dry bulk carriers, but the overall capacity of shipbuilding industry still far left behind by the existing national new building and repair market.

3.3 AVAILABLE MARKET SIZE

Comparing the projected market demand and the existing capacity of the national shipbuilding industry there are approximately 6 to 7 million GT of ship repair market and 400,000 GT new building market have not been catered yet nationally. These numbers are only for the merchant ships, there are also other types of vessels in great demand such as fishing vessels, work boats, and leisure boats etc.

4 GEOGRAPHICAL CONDITION

4.1 THE LOCATION

The proposed location for the integrated shipbuilding cluster is in Tanggamus Regency in the southwest part of Lampung Province, facing Sunda Strait and Banten Province (Java Island) in the south, and surrounded by other regencies. The area available for the cluster is around 5,000 hectares facing to the Semangka Bay [9].

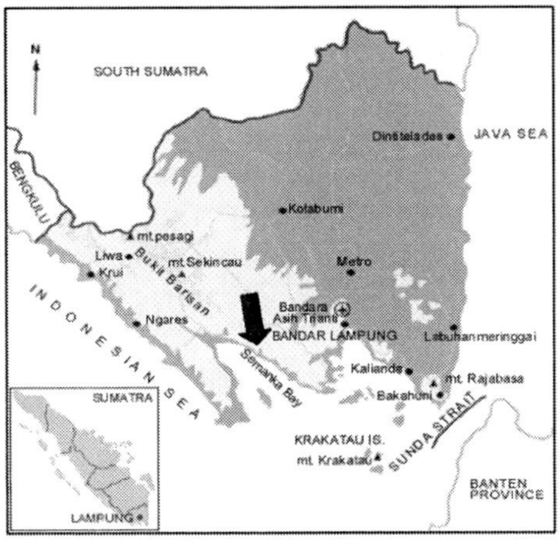

Figure 2: Location Map

Tanggamus Regency has several outstanding natural resources such as geothermal (being explored for electric power), coal and other mines, agro-industries such as palm oil, tapioca, coffee, rubber etc. The regency also passed through by inter provincial road from west part of Sumatra down to Java Island connected by Indonesia busiest ro-ro ferries crossing lane.

4.2 SEA CONDITION

The sea condition of Semangka Bay is very calm and clear because it is protected by Tabuan Island at the entrance to the bay from Sunda Strait and surrounded by rocky coastal line. Sedimentation from the land is very little due to only few rivers flow to the gulf of Semangka. It has depth of around 100m, 5km wide and 15km length [10].

4.3 LAND CONDITION

There is flat rocky land along the sea shore around 100m from the shore line and the land contour is then rise to form a hilly ground. Abrasion by the sea water and erosion from the land are very little. The land is sparsely populated and cultivated by the inhabitant with coconut, cassava, banana and other plantations. There are also small road along the shore and a small fishermen village and fishery harbour exists in the north end of the bay.

5. ECONOMIC DEVELOPMENT IN THE REGION

Economic development of the regency is very much influenced by the development in the region both in Sumatra as well as in Java. In Sumatra the development is mainly dominated by mining, coal in particular, and agro industry such as crude palm oil, tapioca, coffee, rubber, and tropical fruits, beside for export these products are also shipped to Java by means of land transports and barges [11]. In Java especially in Banten province the development is dominated by big scale of chemical, steel mill, construction and heavy industries, and coal generated power plant. The raw materials for the industry in Banten are mainly come from outside the province including Sumatra that mostly shipped by means of sea transports [12].

The great number of land transports (approximately 3,000 trucks, 3,000 buses and cars every day) connecting Sumatra and Java makes ferry crossing lane between Port of Merak and Port of Bakauheni the busiest crossing lane in Indonesia served by around 30 ro-ro ferries every day [13].

6. PROPOSED SHIPBUILDING CLUSTER

Refer to the possible market size and the geographical condition combined with economical development in the region it is proposed that the shipbuilding cluster to be established in Tanggamus Regency would be as follows:

6.1 CORE INDUSTRY

The core industry is the shipbuilding industry which would consist of new building yards, repair yards, scraping or recycle yards, and boat yards. If it is assumed that 10% of the repair market and 15% of the new building market could be grasped from the national available market size there would be 700,000 GT repair works and 60,000 GT new building works are targeted to be carried out each year, and if 1% of the existing fleet to be scraped there would be 170,000 GT for scraping works, combined with the demand for various type of boats the composition of the shipbuilding industry is proposed as follows:

- 1 large repair and new building yard with capacity of 50,000 to 150,000 DWT;
- 1 middle size repair and new building yard with capacity of 15,000 to 50,000 DWT;
- 2 small size repair and new building yards with capacity under 15,000 DWT;
- 1 scraping yard with capacity up to 50,000 DWT;
- 2 boat yards with capacity up to 50 m long.

6.2 PULLING INDUSTRY

The pulling industry is shipping industry and ship owners that give jobs to the shipyards. Beside the national available market that could be grasped by the core industry for the purpose of the industrial cluster there should be commitment from the prospective ship owners and shipping industry to support to reach the goals of the shipbuilding cluster. The main targets from the pulling industry are those that have activities in the region such as the crossing ferries, tankers that do transhipment in Semangka Bay, bulk carriers and barges that supply coal for Suralaya power plant in Banten, chemical tankers and cargo vessels that carry raw materials and products to and from the industries in Banten as well as in Lampung or even Sumatra, fishing and recreational vessels that operate around Sunda Strait.

6.3 SUPPORTING INDUSTRY

Supporting industry is the industry that produces main materials and components for shipbuilding that need further fabrication or work by the shipyards such as steel mill, fabrication industry, freshwater, gas, electrodes and electricity suppliers etc. It is recommended that the supporting industry should be located in the cluster region or not too far from the region for reliability and cost efficiency reasons. In conjunction with this there should be a special area allocated in the cluster, beside there are some industries that have already existed in Banten Province such as PT Krakatau Steel that produces steel plates and sections, and some fabrication industry that would become supporting industry.

6.4 SUPPLYING INDUSTRY

Supplying industry is those that supply all kind of machineries and equipment for the ship that only need to be installed or applied by the shipyards including paint, chains, zinc anodes, navigation and communication equipment etc. It is much better if the supplying industry is also located in the cluster region, but since some of the supplied machineries and equipment are imported from abroad and their sizes and number are considerably shippable the location of the supplying industry can be compromised dependent on the situation as long as there is good access to the cluster's location and high commitment from the suppliers to the successful implementation of the project.

6.5 SUPPORTING SERVICES

Supporting services are institutions that provide services both directly and indirectly to the shipbuilding industry such as classification societies, design and surveying firms, educational institutions, financial institutions, marine consultants etc. These services should have good access to the cluster and commit themselves to the successful implementation of the project.
The cluster should facilitate for the operation of the supporting services.

6.6 INFRASTRUCTURE

In order to foster the development of the shipbuilding cluster the local government should provide infrastructure and supplies to the location of the cluster. Among other things the infrastructure that should be provided are the access roads, if possible special rail ways and port, electricity, freshwater, communication facilities, social and public facilities etc.

Refer to the geographical condition of the proposed location the required infrastructure can be readily developed due to the following reasons: availability of geothermal power plant, rivers that flow near to the location, provincial road that pass through the location, deep and calm water front, flat and rocky land shape etc.

7. CONCLUSIONS

Refer to the above considerations it is concluded that:

From the market opportunity aspect there are approximately 700,000 GT repair works and 60,000 GT new building works, and 170,000 GT for scraping works can be targeted to be carried out each year in the proposed shipbuilding cluster.

From the geographical aspect Tanggamus Regency in Lampung Province has deep, calm, and clean sea water front, has good land condition, and lowly populated land that is very suitable to be developed for shipbuilding industry.

From the commercial aspect Tanggamus Regency in Lampung Province is surrounded by mining and agro-industry on Sumatra side, chemical and heavy industries on Java side, and busiest ferry crossing lane which, are potential for the shipbuilding market.

Thus from the overall point of view Tanggamus Regency in Lampung Province is suitable to be developed as shipbuilding cluster which, proposed to have 1 large, 1 middle size, 2 small size ship repair and new building yards, 1 scraping yard, and 2 boat yards as core industry; ro-ro ferries, tugs and barges, oil and chemical tankers, container vessels, fishing vessels, work and leisure boats that are operating around Lampung Province as the pulling industry; steel mill, fabrication industry, gas, fresh water, and electricity plant in the region as the supporting industry; available supplying industry and supporting services.

ACKNOWLEDGEMENTS

Author would like to thank IPERINDO (Association of Indonesia's shipbuilding and offshore industry) and INSA (Indonesia's national ship owners association) for their contribution of data and information, and the Ministry of Industry for its support in establishing the shipbuilding cluster in Lampung Province.

REFERENCES

1. SECRETARY OF STATE REPUBLIC OF INDONESIA, 'Presidential Instruction No.5 – 2005 on the Empowerment of National Shipping Industry' (*Indonesian*), Secretariat Of State Republic Of Indonesia, 2005.
2. ANAM SAIFUL, 'INSA Asked Businessmen to Invest in National Shipbuilding Industry' (*Indonesian*), Indonesia Shipping Times, 2012.
3. JIBI, 'The Competitiveness of National Shipbuilding Industry And Shipyards Are Low' (*Indonesian*), Bisnis Indonesia, 2010.
4. SAN DIEGO REGIONAL TECHNOLOGY ALLIANCE, 'What Are Industrial Cluster', *San Diego Association of Governments*, 1995.
5. DIRECTORATE OF MARITIME INDUSTRY AND TECHNOLOGY, 'Development Application Concept of Shipbuilding Industry Cluster' (*Indonesian*), Ministry of Industry Republic of Indonesia, 2006.
6. HIDAYAT SOFYAN NUR, '5,000 Ships need routine repair every year' (*Indonesian*), Kontan Newspaper, 2011.
7. PUSDATIN, 'Shipyard Capacity Exceeds Target' (*Indonesian*), Ministry of Industry Republic of Indonesia, 2012.
8. HARYANTI DINI, 'National Shipbuilding Industry' (*Indonesian*), Jurnal Nasional Newspaper, 2012.

9. BAPPEDA, 'Lampung Province's Strategic Development Planning' (*Indonesian*), Lampung Province Government, 2011.
10. BAPPEDA, 'Atlas of Lampung Province' (*Indonesian*), Lampung Province Government, 2011.
11. BPS LAMPUNG, 'Industrial Produces of Lampung Province', Lampung Province Statistical Bureau, 2011.
12. BPS BANTEN, 'Industrial Produces of Banten Province', Banten Province Statistical Bureau, 2011.
13. HUTABARAT CHRISTINE, 'Congestion in Port of Merak' (*Indonesian*), PT ASDP Indonesia Ferry, 2012.

AUTHOR'S BIOGRAPHY

Sunaryo holds the current position of Head of Naval Architecture Study Program and Associate Professor at University of Indonesia. He is responsible in managing the curriculum and educational activities of the Study Program, and delivering lecture on the subject of Ship Production Management. His previous experience includes conducting sabbatical research on Ship Production Management at Strathclyde University in 2010, advisor to the Association of Indonesia's Shipbuilding and Offshore Industry, consultant to the Directorate of Maritime Industry and Technology ministry of Industry Republic of Indonesia.

RUDER CONTROLLED DESIGN USING MODEL REFERENCE ADAPTIVE CONTROL (MRAC) BY OPTIMALIZED REFERENCE MODEL

J.M Herman, Marine Engineering, Pattimura University

SUMMARY

This paper talk about rudder controlled to minimized error because of waves and streams. That error is done to minimized fuel consumption. That things is done by using MRAC (Model Reference Adaptive Control). Adaptive law is done by using Liapunov Method to ensure the system stability. Mode Parameter controller, K_{pm} and K_{dm} are found by using Linear Quadratic Optimal Control to minimise the energy consumption. To damped high frequency, a low pass filter is used so rudder position still fixed although control signal is oscilating.

1. INTRODUCTION

The reason of topography makes a ship sails not on a straight line but it sails in the zigzag line as shown in figure 1. From a departure to a destination a ship will sail thorough some way points. Because of the way points an autopilot has to act like course keeping and course changing. There are many different methods for designing such a controller. In this paper I use Model Reference Adaptive Control (MRAC).

The idea of Model Reference Adaptive Control is to create a closed loop controller with parameters that can be updated to change the response of the system to match a desired model. The model that used in this paper is made based on optimal model.

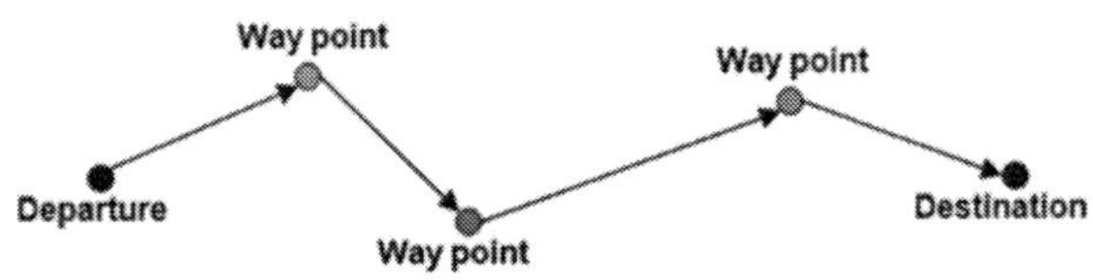

Figure.1. Way points between the departure and the destination.

2. ADAPTIVE LAW

By Nomoto's ship dynamics equation, and by adding Kw as an added input, Nomoto's ship dynamics equation can be written as

$$\tau\ddot{\psi} + \dot{\psi} = K\delta + K_w \quad (1)$$

Where: τ = time constant of a ship. (sec.)
ψ = ship heading (yaw angle) (deg.)
$\dot{\psi}$ = yaw angular velocity or rate of turn (deg./sec.)
$\ddot{\psi}$ = yaw angular acceleration (deg./sec.^{2})
K = yaw rate gain (1/sec.)
δ = rudder angle (deg.)
K_w = disturbances

- By using control signal [Thor I. Fossen [1994], *Guidance and Control of Ocean Vehicles*, John Wiley & Sons, England, p.284]

$$\delta = K_p(\psi_r - \psi) - K_d\dot{\psi} - K_i \quad (2)$$

where:
δ is control signal,
K_p is proportional gain of the system,
K_d is differential gain of the system,
K_i is integral gain of the system,
ψ_r is input reference of the system.

the system can be described in state space form as

$$\dot{X}_p = A_c X_p + B_c U \quad (3)$$

where: $X_p^T = (\psi, \dot{\psi})$ $\qquad U^T = (\psi_r, 1)$

$$A_c = \begin{bmatrix} 0 & 1 \\ -\frac{K K_p}{\tau} & -\frac{(1+KK_d)}{\tau} \end{bmatrix}$$

$$B_p = \begin{bmatrix} 0 & 0 \\ \frac{K K_p}{\tau} & (K_w - KK_i)\frac{1}{\tau} \end{bmatrix}$$

The needed performance of the system can be found from reference model

$$\dot{X}_m = A_m X_m + B_m U \quad (4)$$

Where:

$X_m^T = (\psi_m \ , \ \dot{\psi}_m)$ $\qquad U^T = (\psi_r, 1)$

$$A_m = \begin{bmatrix} 0 & 1 \\ -\frac{K_m}{\tau_m} & -\frac{1}{\tau_m} \end{bmatrix} \qquad B_m = \begin{bmatrix} 0 & 0 \\ \frac{K_m}{\tau_m} & 0 \end{bmatrix}$$

$$K_m = \frac{KK_{pm}}{1+KK_{dm}} \qquad \tau_m = \frac{\tau}{1+KK_{dm}}$$

By taking error as $e = X_p - X_m$, we find,

$$\dot{e} = A_m e + (A_c - A_m) X_p + (B_c - B_m) U$$

$$= A_m e + a\dot{\Psi} + b\varepsilon + c.1$$

where:

$$c = \frac{1}{\tau_m} - \frac{1 + KK_d}{\tau}$$

$$b = \frac{KK_p}{\tau} - \frac{K_m}{\tau_m}$$

$$\varepsilon = \Psi_r - \Psi$$

$$c = (K_w - KK_i)\frac{1}{\tau}$$

Normally a control signal in MRAC is formatted as [Hans Butler [1992], *Model Reference Adaptive Control*, Prentice Hall International, UK, p.8]

$$\delta = \theta^T \omega$$

Where:
$\theta =$ the controller parameter vector.
$\omega =$ a signal vector.

By signal control $\delta = K_p(\Psi_r - \Psi) - K_d\dot{\Psi} - K_i$, we found:

$$\delta = \theta^T \omega = [K_p \quad K_d \quad K_i]\begin{bmatrix} \Psi_r - \Psi \\ -\dot{\Psi} \\ -1 \end{bmatrix}$$

So, $\theta^T = [K_p \quad K_d \quad K_i]$ and $\omega^T = [\Psi_r - \Psi \quad -\dot{\Psi} \quad -1]$

The Liapunov function that chosen is a function that consist of signal error $e = X_p - X_m$ and parameter error vector, $\Phi = \theta - \theta^*$. Vector θ is the vector parameter controller, and θ^* is the vector parameter model. In most simple form, the Lyapunov function is selected as [Hans Butler [1992], *Model Reference Adaptive Control*, Prentice Hall International, UK, p.12]

$$\boldsymbol{V = e^T P e + \Phi^T \Gamma^{-1} \Phi} \qquad (5)$$

To be acceptable as the Lyapunov function candidate the matrices P and Γ^{-1} must be positive definite. Matrix Γ is called the adaptation gain matrix.

Differentiated V, $\dot{V}$, must be negative definite in order to obtain an asymptotically stable adaptive system [Hans Butler [1992], *Model Reference Adaptive Control*, Prentice Hall International, UK, p35]
Differentiating V is

$$\dot{V} = -e^T Q e + some\ terms\ including\ \Phi$$

By putting the extra terms including Q to zero and if Q is positive definite, $\dot{V}$ is guaranteed to be negative definite with respect to e.
Lyapunov's theorem states that for any asymptotically stable system governed by matrix A, a positive symmetric matrix Q yields a positive symmetric matrix P by the equation
[Hans Butler [1992], *Model Reference Adaptive Control*, Prentice Hall International, UK, p.39]

$$\boldsymbol{A^T P + PA = -Q} \qquad (6)$$

By this Lyapunov's equation matrices P and Q can be selected.

We see that

$$\boldsymbol{\dot{V} = e^T (A_m^T P + PA_m)e + 2e^T PB_I \Phi^T \omega + 2\Phi^T \Gamma^{-1} \dot{\Phi}} \quad (7)$$

By putting the last two terms of equation (7) to zero, we find

$$2e^T PB_I \Phi^T \omega + 2\Phi^T \Gamma^{-1} \dot{\Phi} = 0$$

$$\dot{\Phi} = -\Gamma^{-1} e^T PB_I \omega$$

or $$\boldsymbol{\dot{\Phi} = -\Gamma^{-1}(P^T e)\omega} \qquad (8)$$

where:
$PB_I =$ a vector consist of the nth column p of matrix P.
$p^T e =$ the compensator error that used in the adaptive law to calculate $\dot{\Phi}$.

By putting some terms including Q to zero we find adaptive law

$$\boldsymbol{\dot{\theta} = -\Gamma \in \xi} \qquad (9)$$

Where: ξ is a modified version of ω.

$\in$ is a linear combination p^T e of the elements of a state error $e = X_p - X_m$

In Model Reference Adaptive System the parameter update law is [Hans Butler [1992], *Model Reference Adaptive Control*, Prentice Hall International, UK, p.13]

$$\boldsymbol{\dot{\theta} = -\Gamma \in \xi = \dot{\Phi}} \qquad (10)$$

By using equation (8) we find feedback gain as

$$\dot{\Phi} = -\Gamma^{-1} e^T PB_I \omega$$

$$= -\begin{bmatrix}\gamma_0 & 0 & 0\\ 0 & \gamma_1 & 0\\ 0 & 0 & \gamma_2\end{bmatrix}[e \quad \dot{e}]\begin{bmatrix}P_{12}\\ P_{22}\end{bmatrix}\begin{bmatrix}\varepsilon\\ -\dot{\Psi}\\ -1\end{bmatrix}$$

$$= (P_{12}e + P_{22}\dot{e})\begin{bmatrix}-\gamma_0\varepsilon\\ \gamma_1\dot{\Psi}\\ \gamma_2.1\end{bmatrix}$$

or

$$\left.\begin{aligned}\dot{K}_p &= -\gamma_0'(P_{12}e + P_{22}\dot{e})\varepsilon\\ \dot{K}_d &= \gamma_1'(P_{12}e + P_{22}\dot{e})\dot{\Psi}\\ \dot{K}_i &= \gamma_2'(P_{12}e + P_{22}\dot{e}).1\end{aligned}\right\} \quad (11)$$

P_{12} and P_{22} can be found from the Lyapunov function

$$A_m^T P + P A_m = -Q \qquad (12)$$

and $\gamma_i' = \frac{\gamma_i}{2}$. The factor 2 comes from the process control vector and is generally not known, which does not complicate the implementation while any $\Gamma > 0$ suffices [Hans Butler [1992], *Model Reference Adaptive Control*, Prentice Hall International, UK, p42]
. γ_i is called adaptation gain.

A_m, P, and Q is calculated based on

$$A_m = \begin{bmatrix}0 & 1\\ -\frac{K_m}{\tau_m} & -\frac{1}{\tau_m}\end{bmatrix} \qquad P = \begin{bmatrix}p_{11} & P_{12}\\ P_{21} & P_{22}\end{bmatrix} \qquad Q = \begin{bmatrix}-1 & 0\\ 0 & -1\end{bmatrix}$$

Where

$$P_{12} = \frac{\tau_m}{2K_m} \qquad P_{22} = \frac{\tau_m^2}{2K_m} + \frac{\tau_m}{2}$$

By the equations that we have done, a block diagram can be made like figure 2.

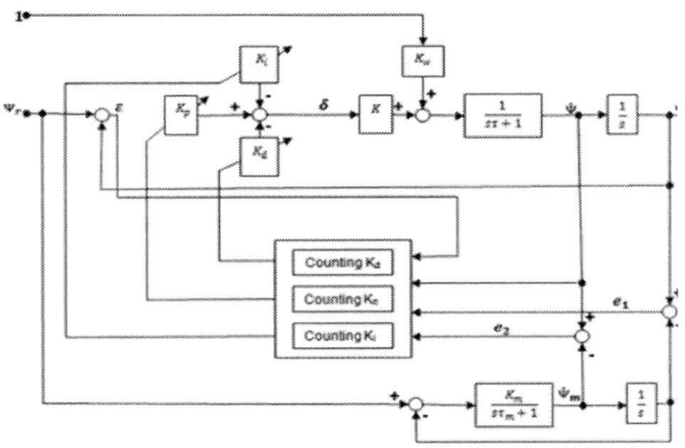

Figure.2. Ship control system that will be made.

Figure 2 is made based on the equations below

$$\tau\ddot{\Psi} + \dot{\Psi} = K\delta + K_w$$

$$\delta = K_p(\Psi_r - \Psi) - K_d\dot{\Psi} - K_i$$

$$\tau\ddot{\Psi}_m + \dot{\Psi} = K\delta_m$$

$$\delta_m = K_{pm}(\Psi_r - \Psi_m) - K_{dm}\dot{\Psi}_m$$

$$e = \begin{bmatrix}\Psi\\ \dot{\Psi}\end{bmatrix} - \begin{bmatrix}\Psi_m\\ \dot{\Psi}_m\end{bmatrix} = \begin{bmatrix}\Psi - \Psi_m\\ \dot{\Psi} - \dot{\Psi}_m\end{bmatrix} = \begin{bmatrix}e_1\\ e_2\end{bmatrix}$$

Adaptive laws are calculated according to equations (11)

$$K_p = -\gamma_0'\,(P_{12}e + P_{22}e)\,\varepsilon$$

$$K_d = \gamma_1'\,(P_{12}e + P_{22}e)\,\Psi$$

$$K_i = \gamma_2'\,(P_{12}e + P_{22}e)\,1$$

K_m, gain of the model, and τ_m, time constant of the model are calculated based on

$$K_m = \frac{KK_{pm}}{1+KK_{dm}} \qquad (13)$$

$$\tau_m = \frac{\tau}{1+KK_{dm}} \qquad (14)$$

K and τ is calculated based on[8]

$$K = \frac{v}{v_0}K_0 \qquad (15)$$

$$\tau = \frac{v}{v_0}\tau_0 \qquad (16)$$

Where:
v is the actual speed of ship
v_0 is the nominal speed of ship

For the model $v = v_0$ so $K = K_0$, and $\tau = \tau_0$.

3. OPTIMIZATION K_{pm} and K_{dm}

To make K_{pm} and K_{dm} optimum we follow the steps below

- Determine criterion function,
- Determine Hamilton function,
- Take partial derivative to δ and x to find stationer condition and costate equation,
- Calculate K_{pm} and K_{dm}.

To find criterion function some aspects have to be concern such as:

- Accuracy of ship direction,
- Accuracy of ship position,
- Minimum fuel consumption,
- Something wrong in steering gear that caused lack of steering gear performance.

It is better if we can combine all criteria above to one criterion function with a known weighting factor. It is a pity that we can't do it because weighting factor depend on navigate situation. A ship sails in offshore or crowded traffic needs accuracy in its heading and position. A ship sails in the ocean needs minimum fuel consumption. In this paper criterion function is calculated based on minimum fuel consumption. Koyama (1967) had observed that ship's yaw can be described as a sinusoid (figure 3).

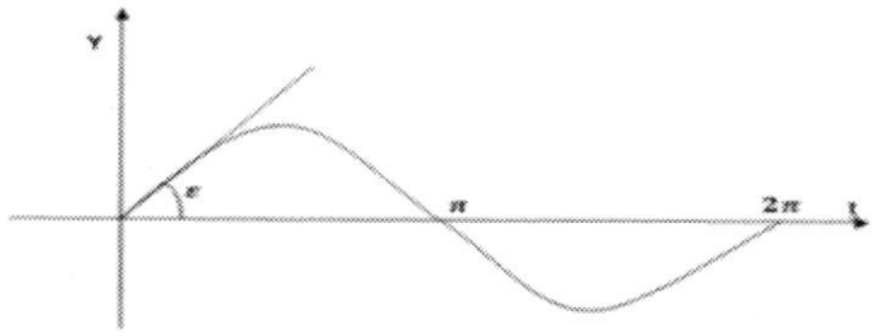

Figure.3. Course error during autopilot

By using elongation in distance Koyama calculated criterion function as [Thor I. Fossen [1994], *Guidance and Control of Ocean Vehicles*, John Wiley & Sons, England, p.266]

$$\boldsymbol{min\,J} = \frac{\mathbf{0{,}0076}}{\boldsymbol{T}} \int_0^T (\,\boldsymbol{\varepsilon^2} + \boldsymbol{R\delta^2}\,)\,\boldsymbol{dt} \quad (17)$$

where: J = loss of speed .... (%)
ε = course error .... (degree)
δ = rudder angle .... (degree)
R = weighting factor

Equation (17) can be written as

$$\boldsymbol{minJ} = \frac{\mathbf{2(0{,}0076)}}{\boldsymbol{T}}\,\boldsymbol{J^*} \quad (18)$$

where: $$\boldsymbol{J^*} = \frac{1}{2}\int_0^T \{(\,\boldsymbol{Cx} - \Psi_r)^2 + \boldsymbol{R\delta^2}\}\,\boldsymbol{dt}$$
(19)

$Cx = \Psi$ = output system (degree)

Ψ_r = input reference (degree)

If J^* is minimum and J also minimum.

From J^* , $\dot{X} = Ax$ and $B\delta$ Hamilton function will be found as

$$\boldsymbol{H} = \frac{1}{2}\int_0^T \{(\,\boldsymbol{Cx} - \Psi_r)^2 + \boldsymbol{R\delta^2}\} + \boldsymbol{\lambda^T(Ax + B\delta)} \quad (20)$$

From equation 20 is found

$$\frac{\partial H}{\partial \delta} = R\delta + B^T\lambda = 0$$

or, $\boldsymbol{\partial = R^{-1}B^T\lambda}$ (21)

where: λ = Lagrange Multiplier.

Lagrange mutiplier is found from costate equation

$$-\dot{\lambda} = \frac{\partial H}{\partial x} = C^T(Cx - \Psi_r) + A^T\lambda$$

or $$\boldsymbol{\dot{\lambda} = -A^T\lambda - C^T(Cx - \Psi_r)}$$
(22)

Although ∂ is a linear costate feedback it can't be expressed as a linear state feedback.
[Frank L. Lewis & Vasilis L. Syrmos [1995], *Optimal Control*, John Wiley & Sons, Inc. Canada, p.217]

By using sweep method { a method assumes that X(t) dan λ(t) fulfill a linear relationship of a unknown function S(t) }

$$\boldsymbol{\lambda(t) = S(t)X(t)} \quad (23)$$

It can be shown that optimal LQ tracker could be expressed in the form [Frank L. Lewis & Vasilis L. Syrmos [1995], *Optimal Control*, John Wiley & Sons, Inc. Canada, p.217]

$$\boldsymbol{K(t) = R^{-1}B^TS(t)} \quad (24)$$

$$\boldsymbol{-\dot{S} = A^TS + SA - SBR^{-1}B^TS + C^TC} \quad (25)$$

where:

$$A = \begin{bmatrix} 0 & 1 \\ 0 & -\frac{1}{\tau} \end{bmatrix} \quad ; \quad B = \begin{bmatrix} 0 \\ \frac{K}{\tau} \end{bmatrix} \quad ;$$

$$S(t) = \begin{bmatrix} S_1(t) & S_2(t) \\ S_2(t) & S_3(t) \end{bmatrix}$$

Calculating steady state of equation 25 we find

$$\boldsymbol{S_2(t)} = \frac{\tau}{K}\sqrt{\boldsymbol{R}} \quad (26)$$

$$\boldsymbol{S_3(t)} = \frac{\mathbf{2}S_2}{\frac{2}{\tau} + \left(\frac{K}{\tau}\right)^2 R^{-1}} \quad . (27)$$

$$\boldsymbol{S_1(t)} = \frac{1}{\tau}\boldsymbol{S_2} + \left(\frac{K}{\tau}\right)^2 \boldsymbol{R^{-1}S_2S_3} \quad (28)$$

From equation 24 is found

$$\boldsymbol{K_{pm}} = \frac{K}{\tau}\boldsymbol{R^{-1}S_2(t)}$$

$$= \sqrt{\frac{1}{R}} \quad (29)$$

$$\boldsymbol{K_{dm}} = \frac{K}{\tau}\boldsymbol{R^{-1}S_1} \quad (30)$$

4. SIMULATION

Simulation was done in two step using MATLAB-SIMULINK program. Step one was done for model and step two was done for the system. Data is taken from Thor I. Fossen [1994], *Guidance and Control of Ocean Vehicles*, John Wiley & Sons, England, p, 174. [L= 161 m; v_0= 7.7 msec^{-1}; K_0= 0.185 sec^{-1}; τ_0= 144.3 sec.]

For the model simulation was done by using block diagram as shown in figure 4.

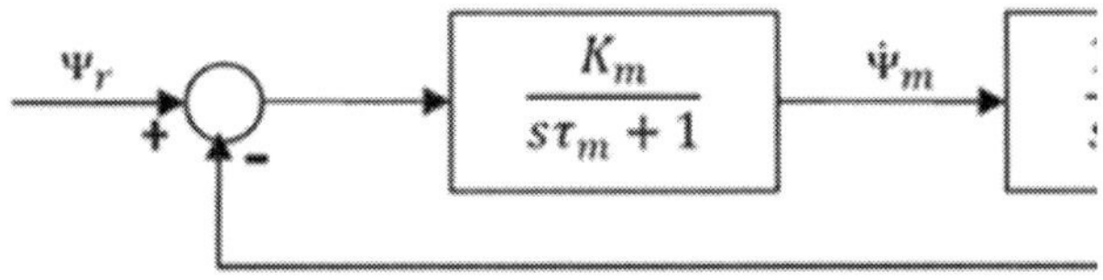

Figure.4. Block diagram of the mode

The result of model simulation is shown in figure 5.

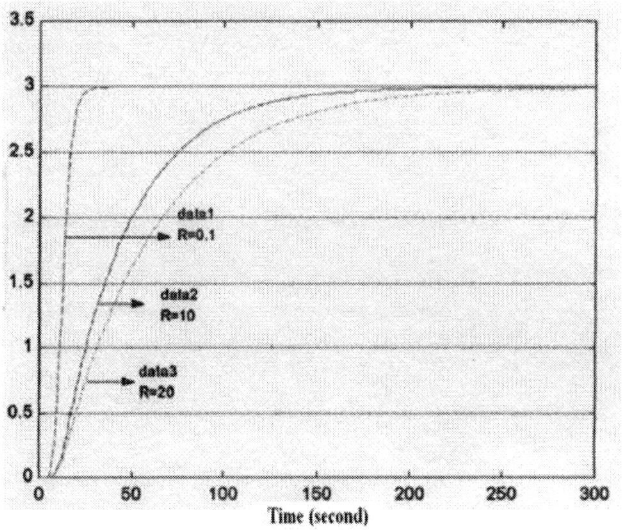

Figure.5. Result of model simulation

Equations 15 and 16 are shown that yaw rate gain K and time constant τ are depend on speed of the ship, v.

Speed of a ship can change according to

- Yaw motion,
- Ship in riding on the wave condition.

Disturbance is calculated according to[10]

$$K_w = 2\xi\omega_0\sigma_w \qquad (31)$$

where:

ξ is damping coefficient
ω_o is wave frequency (rad./sec)
σ_w is wave intensity

The wave frequency can be calculated by[11]:

$$\omega_o = 2\pi f_0 \qquad (32)$$

where:

f_0 is modal frequency (cec^{-1})

For sailing ship in v speed the wave frequency can be modified as[12]:

$$\omega_e = \omega_o - \frac{\omega_o^2}{g} v \cos\Omega \qquad (33)$$

where:

ω_o is frequency of the wave hitting the ship (rad/sec)
Ω is angle of the wave hitting the ship (rad)
g is gravitation (m/sec.^{2})
v is speed of the ship (m/sec.)

Because of ω_e we can modify K_w in the equation 31 as:

$$K_w = 2\xi\omega_e\sigma_w \qquad (34)$$

$K_w = f(v, \Omega, \omega_o, \sigma_w)$. It is shown in equation 33 and equation 34.
Waves produce large oscillator forces and moments. They will cause to much control action. To avoid high frequencies rudder motion we need waves filtering. There are some filters. In this paper, a low pass filter because it can suppress high frequencies rudder motion.
Simulation is done by step function setting point as input with differences K_w. Results are shown in figure 6 to figure 9.

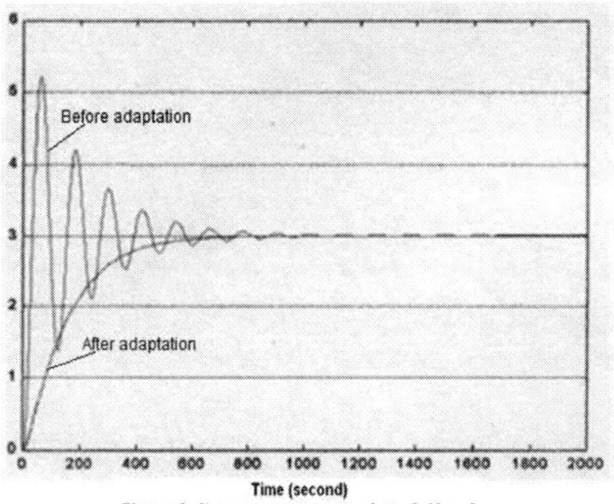

Figure.6. Output system; set point = 3, K_w = 0

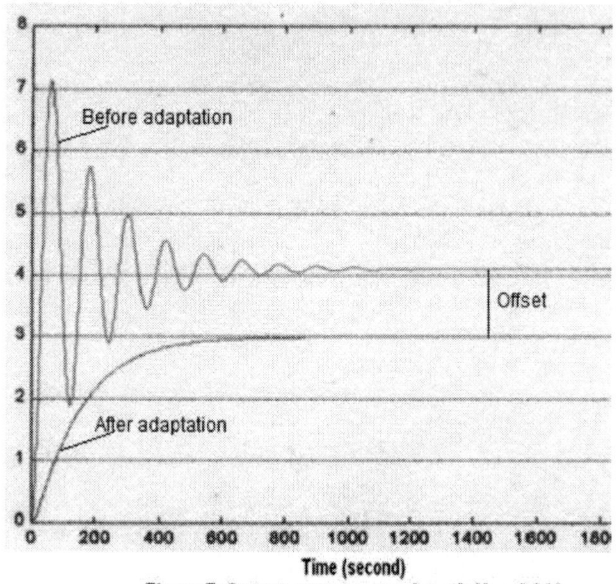

Figure.7. Output system; set point = 3, K_w = 0.241

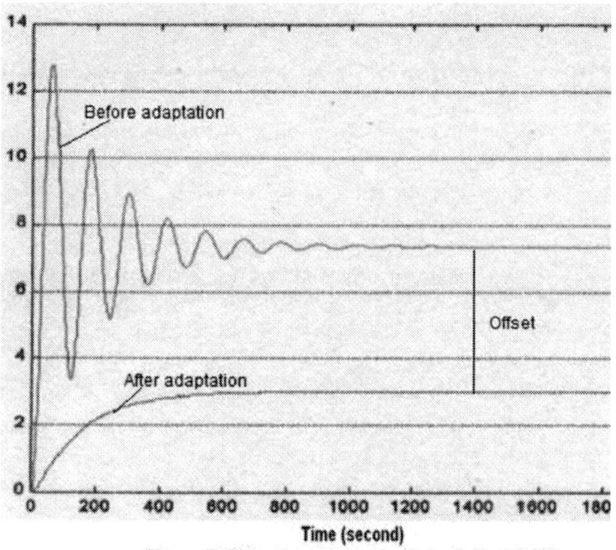

Figure.8. Output system; set point = 3, K_w = 1.186

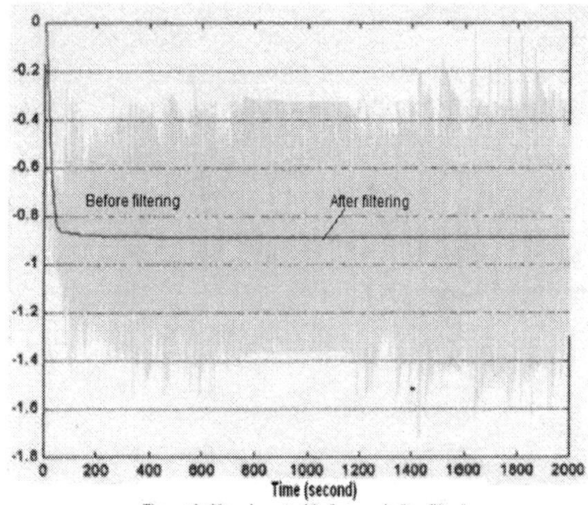

Figure.9. Signal control before and after filtering

5. CONCLUSION

It can be shown in figure 6 to 8 that the rudder can be controlled perfectly. It can be said because output of the system can follow the input perfectly. It can be seen in figure 6 that in the no disturbance condition there is not an offset. In the disturbance condition there is an offset as shown in figure 7 and 8. The offset is depends on the disturbances. It is shown in figure 9 that the low pass filter using to reduce high waves frequencies also works perfectly.

6. REFERENCES

- Brian D.O Anderson & John B. More [1989], *Optimal Control, Linear Quadratic Methods*, Prentice Hall International, USA.
- C. W. Merriam III [1964], *Optimization Theory and Design of Feedback Control System*, McGraw Hill, Inc.USA.
- Frank L. Lewis & Vasilis L. Syrmos [1995], *Optimal Control*, John Wiley & Sons, Inc. Canada.
- Hans Butler [1992], *Model Reference Adaptive Control*, Prentice Hall International, UK.
- Karl Johan Åström, Björn Wittenmark [1995], *Adaptive Control*, Addison Wesley Publishing Company, USA.
- Raymond T. Stefani [1994], *Design of Feedback Control System*, Saunders College Publishing, USA.
- Thor I. Fossen [1994], *Guidance and Control of Ocean Vehicles*, John Wiley & Sons, England.

CONCEPT SELECTION OF INTER-ISLAND TRANSPORT SYSTEM AT THE SUNDA STRAIT

D.M Rosyid, Department of Ocean Engineering, ITS Surabaya, Indonesia

SUMMARY

The government of Indonesia is considering to build an inter-island bridge across the Sunda Strait located between Java and Sumatra. This initiative was triggered by a need to accelerate regional development in Sumatra, and the poor performance of the present ferry service. This paper attempts to develop a concept selection approach to inter-island transport system using a topology approach. A risk-based and techno-economic approaches to concept selection are also presented. We concluded that an advanced ferry system is superior to a bridge system.

1. INTRODUCTION

The government of the Republic of Indonesia has recently launched the Master Plan Percepatan Pembangunan Ekonomi Indonesia (MP3EI) or the Master Plan for Accelerating Economic Development of Indonesia. The MP3EI is broken down into 6 development regional corridors each focusing on some particular regional strength of the corridor. In the eastern part of Indonesia, the Maluku-Papua corridor for example will be based on fisheries and maritime economics. The Morotai island on the north of Maluku is planned to provide an international terminal facilities to support international shipping through the Indonesia Sea Lane III [1].

In the western part of Indonesia, the Java-Sumatra corridor of economic development will rely upon major infrastructural investment to support a regional economy based on agricultural resources such as palm oil in particular and manufactured products. Within this scenario, a big economic push will be provided by a major inter-island transport infrastructural investment at the Sunda Strait.

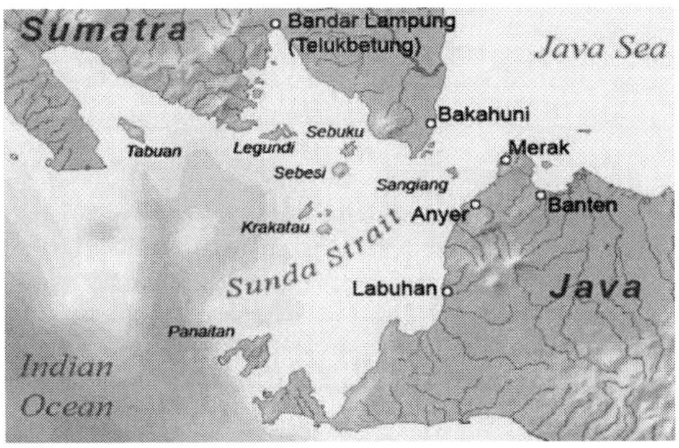

Figure 1: Merak-Bakahuni at The Sunda Strait

The idea of building an inter-island bridge "connecting" Java and Sumatra has been around for quite some time, even during the Soekarno era in early 1960-s. Only until at the beginning of the 21st century the idea gets significant support by the present government. This is partly due to the poor performance of the existing ferry services connecting Merak at the Java side (Banten Province) to Bakauheni at the Sumatra side (Lampung Province). This paper will attempt to provide a more careful investigation on the viability of the idea.

First of all, it must be said that the poor performance of the Merak-Bakauheni ferry service is not unique, i.e. generally speaking, all ferry services in Indonesia is not something many Indonesian can be proud of. The services are generally provided by outdated ferry design, second-hand, poorly maintained fleet [2]. Almost no operators are capable of raising funds to purchase ferries of new, and latest generation design. With the most recent ferry accidents in the Sunda strait in late September 2012, the poor safety performance of the ferry operating in the strait is further confirmed.

The generally poor performance of Indonesian ferry results from a lack of support of the transport authority and the banking community to provide better investment climate that may enable for ferry operators to provide newly built ferries of the latest design. The case for the domestic shipping industry is very much similar. This has led to poor safety and efficiency record of the national fleet. Most ferry terminal and other shore infrastructures in Indonesia are also poorly developed and maintained.

However, this paper will argue that comparing the poorly performing existing ferry services against a newly proposed bridge is not fair. The comparison should have been between a new bridge to a new ferry service system. The publicly disclosed proposed bridge is of a suspension bridge technology divided into 7 segments with a total length of approximately 30 km. The segments of the bridge are to be based upon several small islands located somewhere between Merak and Bakauheni. The longest span of the bridge is approximately 3000m. This alone will be a world record. The height of its supporting pylon will be at least 80m above sea level, at the water depth of at least 40m. It will accommodate 6 traffic lanes, without a double rail track. The cost of constructing the bridge is estimated at Rp.

180-200 T. It will take about an estimated 10 years to build.

The design and construction challenge of the bridge are both well beyond the international experience. The sea current at the Sunda Strait is notoriously strong, and both the seismic and volcanic activities of the surrounding areas are very active. The Krakatau volcanic system is one of the most active in the world. Every technology that challenges directly against the force of nature will surely cost dearly if technically feasible.

On the other hand, the design and construction challenge of a comparable, advanced ferry system are well within international experience. The proposed, competing ferry system can be specified easily by assessing the present fleet size operating at the Sunda Strait. We estimate that an advanced ferry system can be made very much cost-effectively and within a much shorter time period. The capacity of the system is also easily adapted to varying future demand in the traffic. This flexibility is a key distinctive feature of the proposed advanced ferry system.

In the following, a topology argument is developed to select between a bridge system and its contending ferry system. Topology is a major area of mathematics concerned with the most basic properties of space, such as connectedness. Topology developed as a field of study out of geometry and set theory, through analysis of such concepts as space, dimension, and transformation. Topology can be defined as the study of qualitative properties of certain objects (called topological spaces) that are invariant under a certain kind of transformation (called a continuous map), especially those properties that are invariant under a certain kind of equivalence (called homeomorphism). To put it more simply, topology is the study of continuity and connectivity [3].

2. INTER-ISLAND TRANSPORT SYSTEM : A TOPOLOGY APPROACH

A bridge system is any structure that connects two sides of land mass separated by a body of water so that one can cross over the water. The body of water may take the form of a river, a gulf, or a strait. A bridge can also be any structure that connects two sides of masses separated by a vertically discontinued space so that the bridge is a shorter trace than a trace following the vertically discontinued space. The bridge connecting the twin towers of PETRONAS is such a bridge.

A perfect circle or rectangle is a convex 2-D space. One small, irregular crack on the circle will make it a concave space. Cutting one small single circular hole of a perfect circle will also make it a concave space. However, a straight cut dividing through a convex space will produce two smaller convex spaces.

From topology point of view, an island is made to be a (horizontally) concave space of land mass domain by the presence of at least a river or a gulf. A concave space is the one in which there exists a straight line connecting a pair of points in the space that lies outside the space. A bridge is such a line in the concave space. A bridge is therefore a shortest path solution in a horizontally concave space.

Two islands separated by a strait does not make a concave space. However, a bridge "connecting" the two islands will create a concave space. Thus, a distance problem will always result from connecting two islands : there exist pairs of points that lines drawn to connect them will lie outside the space.

Therefore, from topology argument, a bridge is a shortest path solution to a distance problem only in a concave (vertically or horizontally) space of land mass domain. A bridge connecting two islands create a new distance problem by creating a previously non-existing concavity. A bridge can be a better solution for a dried strait without the seawater.

A ferry system consists of a fleet of ferry, a terminal subsystem at each side of the island, and the sea water subsystem. The sea water subsystem decreases the degree of vertical concavity. A trace on the seawater is always shorter than any trace following a vertically discontinued space of the seabed. The seawater is a natural, and unlimited number of bridges for the fleet to enjoy. The seawater subsystem of a ferry system renders its horizontal connectedness of the two islands much more superior than the bridge system. In fact, a bridge system is less connected than a ferry system.

From topology argument we may conclude that a ferry system does not create a new distance problem while a bridge system does.

3. RISK-BASED SYSTEM CONCEPT SELECTION

The competing concepts of inter-island transport system are now compared in terms of the risks the two may be exposed to. The following hazards can be identified and their corresponding risks are qualitatively estimated. (See Table 1 and Table 2 below). The design failure (DF) of the bridge is much more probable than the design failure of the ferry system. The bridge design has to face a larger degree of uncertainty posed by the seabed dynamic characteristic, while the ferry design is not exposed to this. The consequence of bridge design failure is catastrophic, while a ferry design failure will only lead to a relatively moderate consequence.

The construction failure (CF) of the bridge is also much probable than a ferry construction failure. The consequence of construction failure of the bridge is of major consequence, while the ferry construction failure is of minor consequence. The environmental impacts (EI) during the construction of the bridge is very likely to happen with a moderate consequence. The ferry environmental impacts is likely to happen with a minor consequence.

Learning from similarly major projects such as the Euro Tunnel [2], the bridge cost over run (COR) is very likely to happen with substantial financial consequences. The ferry cost overrun is unlikely to happen with minor consequence. Completion delays (CD) of the bridge is very likely to happen with moderate consequences. Completion delays of ferry is unlikely to happen and its consequence is minor. From a business failure perspective (BF), the business of operating the bridge is very likely to loss rather than profit. Its consequence is moderate. With a substantially lower investment involved, operating the ferry is unlikely to loss with minor consequence.

The bridge is like a lame duck and therefore an easier target for an unlikely terrorist attack (TA). Its consequence is minor. A ferry is a flexible system due to its mobility and a more difficult target for a remote attack and its consequence is negligible. From operation (OF) point of view, operating the bridge to be useable and safe is much more difficult due to its size, and its challenging environment. The consequence is moderate. Operating a ferry system is unlikely to pose any major problem, and its consequence is minor.

From a negative regional economics point of view, the bridge construction may be likely to benefit skilled workers only, and this means a limited employment opportunity to unskilled worker. Land overtake by people from the stronger side of the strait will lead to poverty to the weaker side. The consequences are serious and far reaching. The risk profile of the proposed bridge and its contending advanced ferry System are shown in Table 1 and Table 2 below :

Table 1. Risk Profile of the Proposed Bridge System

Degree of Likelihood	Degree of Consequence				
	Negligible	Minor	Moderate	Major	Catastrophic
Extremely Likely					
Very Likely		Completion Delays CD	Cost overrun COR	Bussiness failure BF	Regional Negative Impact RNI
Likely			Operation failure OF	Environmental Impact EI	Construction and Design Failures (DCF)
Unlikely		Terrorist attack TA			
Remote					

Table 2. Risk Profile of the Proposed Advanced Ferry System

Degree of Likelihood	Degree of Consequence				
	Negligible	Minor	Moderate	Major	Catastrophic
Extremely Likely					
Very Likely					
Likely					
Unlikely	RNI	CD/OF/COR	EI/BF/DCF		
Remote	TA				

From the risks profile of the competing systems which we can see qualitatively at concept selection stage, we may conclude that the advanced ferry system is superior in terms of risk that may occur during the design, construction, and operation of the systems.

4. TECHNO-ECONOMIC SYSTEM CONCEPT SELECTION

Indonesia has for some time clearly fallen into a single-mode, i.e. road, trap. Worse, it has drawn itself into an unsustainable, private, fossil fueled road-based transport. For an archipelago country, this is clearly a mistaken solution. The transport policy has in fact been dictated by the public work policy in which road solution dominates at the expenses of other modes of transport. This has undermined the reliability of its national transportation system. River and railway transport have consistently degraded during the last 20 years. Ferry transport is also lacking of support. The initiative to build an inter-island bridge is the continuation of this mistaken policy direction [4]. Within that context, in addition to a risk-based evaluation previously discussed, Table 3 below shows a qualitative, techno-economic evaluation of the competing systems [2].

Table 3. Qualitative comparison between competing systems

Selection Criteria	Bridge System	Advanced Ferry System
Construction Cost	Rp. 200 T	Rp. 10 T
Project Completion	10 years	3 years
Operation and maintenance cost	Substantial	Manageable
Potential cost over run	High	Low
Design uncertainties	Significant	Manageable
Multi-modality promotion	Poor	Very good
Flexibility against future traffic demands	Poor	Very good
Regional Benefit	Significant	Limited
Employment creation opportunities	Significant	Limited
Fairness to nation-wide tax payers	Unfair	Fair
Sensitivity to future financial shocks	Very sensitive	Not so sensitive

From Table 3 above we may conclude that the advanced ferry system is still a better option to select.

5. CONCLUSIONS

Concept selection problem is strongly featured by insufficient quantitative information. A qualitative approach is therefore much more suitable. A topology approach to concept selection of a major inter-island transport system has been presented. A risk-based and techno-economic evaluation have also been presented. The topology approach and techno-economic analysis have clearly indicated that a ferry system is a much more viable option to select.

6. REFERENCES

1. ROSYID, D.M. 'Quality Growth ?'. *Proceedings of a National Seminar on Development, Growth, and Equity*. AKAPSI, Jakarta, October 15, 2012

2. ROSYID, D.M., 'The Archipelago Paradigm for Indonesia Development'. *Professorship Oration*. ITS. October 2010

3. ROSYID, D.M., 'Concept Selection of a Strait Crossing Transport System : a Topology Argument'. *The 1st APTECS Conference*, October 2010

4. ROSYID, D.M., 'Formulating a New Development Paradigm in the 21st Century based upon Local Wisdom'. *Proceedings of the National Seminar on Technology and Social Change*. Department of Humanity Studies. ITS, October 2012.

7. AUTHOR'S BIOGRAPHY

Daniel Mohammad Rosyid received his Ph.D in Marine Technology from the University of Newcastle upon Tyne, UK, in 1991. He is now professor in Operation Research and Optimization at the Department of Ocean Engineering of ITS Surabaya. He is a practicing naval architect with a focus on small island ferry transport. Daniel has been involved in policy studies during the last few years at the Public Administration Graduate program of the Hang Tuah University focusing on marine-related policies.

SEAKEEPING ANALYSIS OF THE TRIMARAN FERRY SHIP IN SHORT CRESTED SEA FOR A CASE OF EAST JAVA WATER CONDITION

Aries Sulisetyono, Department of Naval Architecture and Shipbuilding Engineering, Institut Teknologi Sepuluh Nopember (ITS) Indonesia

SUMMARY

Seakeeping analysis is an important aspect in the assessing of ship performance in wave corresponding to the safety and comfort of ship's crews or passengers. This paper consider to the six degree of freedom (6 DOF) motions of the trimaran ferry ship that was operated in short crested wave of East Java sea, Indonesia. There were some snap shot data representing one year sea condition with of an hour period of snap shot. Based on the snap shot data, the wave spectrums were further generated using the standard ITTC formulation of JONSWAP spectrum. The 3D diffraction analysis was performed to obtain the first-order (high frequency) response motions of ship in regular wave for any ship's heading relative to the waves. The ship's motion responses in multidirectional seas were computed by multiplication of the Response Amplitude Operator (RAOs) and multidirectional wave spectrum. The standard deviations of each motion were obtained by square root of the summation of the total spectral energy of wave. The periods of ship motion were also determined by the division of the total spectral energy to the total spectral moment. The periods and standard deviations were plotted in the scatter charts representing the motion responses for all snap shot data. The cumulative exceeding probabilities of motion were determined, and it was used to examine the response of motions to exceed criteria value.

1. INTRODUCTION

In Sumenep, Madura Island, Indonesia, the demand of passenger and goods transport facilities for inter-island shipping is growing up. Unfortunately, the existing ferry ship was having limitation performance of speed and seakeeping due to a high ocean wave condition. In this case, the trimaran hull design concept was proposed as an alternative solution to challenge the both issues above.

Recently, the trimaran ship design was developed due to it's a good seagoing performance with significant advantages to other ship such as low resistance at high speed which affect to the fuel consumption efficiency [3]; a good stability due to the suitable arrangement of the side hulls [4]; having a large open decks influence to the transport capacity and deck arrangement. The design of high-speed ships, the trimaran has become a good seakeeping performance vehicle type in a high sea [1].

This paper consider to the analysis of seakeeping performance of a trimaran vessel in the short crested seaway for a route case of the Sumenep Madura Island going to the Kangean Island. The new approach was proposed to examine the motion responses of vessel while was operated on the environmental wave condition of the Sumenep – Kangean route. The first-order (high frequency) motion responses of the vessel in regular wave were analyzed using 3D Diffraction analysis or Panel method [7]. The required input data ranges consisted of the vessel main dimensions, vessel geometry, mass and mass moment of inertia, hydrodynamics coefficients, and wave headings. The advancing speed of trimaran vessel was taken for maximum speed value.

In numerical modelling, the hull geometry was panellized into certain number of panel to obtain reasonably accurate prediction. The output data was given in the form of tabulated RAO (Response Amplitude Operator) of the six-degree of freedom (6-DoF) vessel motions, namely surge, sway, heave, roll, pitch, and yaw as a function of wave frequencies (wave periods).

The environmental data that were provided by the Agency of Meteorology, climatology, and geophysical (BMKG) of Indonesia Government, with totally numbers of 8760 cases were selected a one-year wave condition with the relation interest of the height, period and direction of wave. Based on the snap shot data, the wave spectrums were further generated using the standard ITTC formulation of JONSWAP spectrum. Directional spreading to the wave energy was considered follow the snap shot data. The form of the spectrum along each direction was differently, and the heading angle of wave direction relative to the direction of ship was conducted.

Ship's motion responses in multidirectional seas were computed with the multiplication of multidirectional encountered wave spectrum obtained from environmental data and the trimaran's motion response in regular wave (RAO). The significant amplitude responses of each motion were obtained by two times the square root of total spectral energy of motion response. The periods of vessel motion were also obtained by the division of the total spectral energy to the total spectral moment of motion response. Furthermore, the cumulative exceeding probabilities of motions were obtained so that the response to exceed certain value can be identified.

2. SHIP MOTIONS IN REGULAR WAVE

On the concept design stages, the general arrangement of Trimaran Ferry Ship (TFS) was designed [5] as presented in Figure 1, and the main particular of this vessel in full-load condition is contained in Table 1.

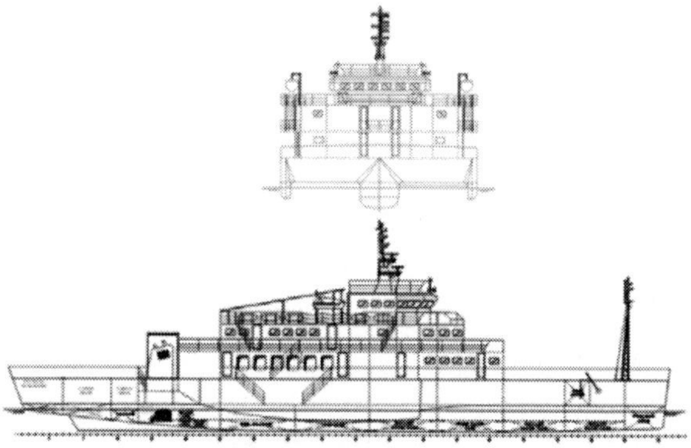

Figure 1 General Arrangement of the Trimaran Ferry Ship

Table 1 Main particular of the Trimaran Ferry Ship

Length of main hull, (Lpp)	54.53 m
Length of side hull, (Lpps)	16.63 m
Breadth of main hull, (B)	3.64 m
Breadth of side hull, (Bs)	0.97 m
Breadth of hull, (Bm)	18.50 m
Depth of main hull, (D)	4.55 m
Depth of side hull, (Ds)	3.70 m
Draft of main hull, (T)	1.82 m
Draft of side hull, (Ts)	0.97 m
Displacement of main hull, (Δ)	187.4 ton
Displacement of side hull, (Δs)	8.06 ton
Centre of gravity above base, (KG)	1.13 m
Centre of gravity from AP, (LCG)	25.38 m
Centre of buoyancy above base, (KB)	1.29 m
Centre of buoyancy from AP, (LCB)	0.82 m
Speed of Ship	16 knot

For seakeeping analysis, the definition of the coordinate system was defined. The coordinate system ***xyzo*** has the origin ***o*** in the vessel centre of mass, *x*-axis positive in the bow direction, the *y*-axis toward port side direction and *z*-axis toward up direction, see Figure 2. The vessel motions were defined in this coordinate system, such as the surge, sway and heave motions were measured along the *x*-axis, *y*-axis and *z*-axis respectively; the roll, pitch and yaw motion were measured about **x**-axis, *y*-axis and **z**-axis respectively.

The vessel hull form was modelled to full scale as a fined mesh with panels representing the hull surface, as depicted in Figure 2. The wave heading angle, μ, relative to the vessel was defined head sea (180^0) while the wave approach to the vessel from the vessel's bow, the following sea (0^0) when the wave coming to the vessel from the stern of vessel, and beam sea (90^0) when wave approach to the vessel from the sides of vessel, see Figure 2. Since the vessel was symmetrical the wave heading angle only consider to 0^0 until 180^0.

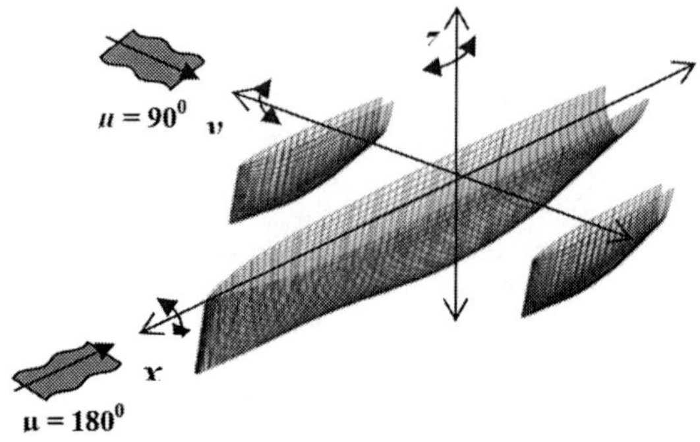

Figure 2 Co-ordinate system and wave heading angle definition

The characteristics of the Trimaran Ferry Ship (TFS) motion in regular wave were calculated by boundary element method (BEM). Wetted surface of the Trimaran was discretized with 2039 number of panels. The hydrodynamics forces were determined by imposing boundary conditions on the wetted surface of the Trimaran in infinity depth water.

The results of RAO (Response Amplitude Operator) for surge, sway, heave, roll, pitch and yaw motion of the Trimaran in regular wave were plotted respectively in Figure 3 through Figure 7. The translation motion such as surge, sway and heave were presented per unit wave amplitude, and the rotation motions of roll, pitch and yaw were presented per unit degree amplitude.

Figure 3, 5 and 7 had shown the surge, heave and pitch motion were having a low amplitude response for all frequency and heading angle. The highest responses of translation and rotation motion were described such as motion of Sway and Yaw respectively for the case of 90 degree heading angle with low frequency, sees Figure 4 and Figure 8.

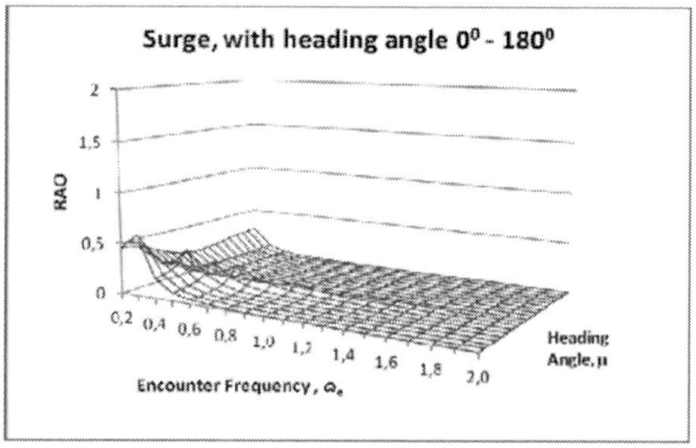

Figure 3 RAO of Surge motion with heading angle 0^0 - 180^0

The rotation response of Roll and Yaw motion which were presented in Figure 6 and Figure 8 respectively, were almost have same response value for difference range of frequencies such as the Roll at high frequency and Yaw at low frequency. And both motions were proceed at the heading angle close to 90 degree.

This phenomenon might be caused by the dimension of main hull breadth which were located on the front, and stern parts small comparing to the middle part of ship, so that the damping coefficient having small value.

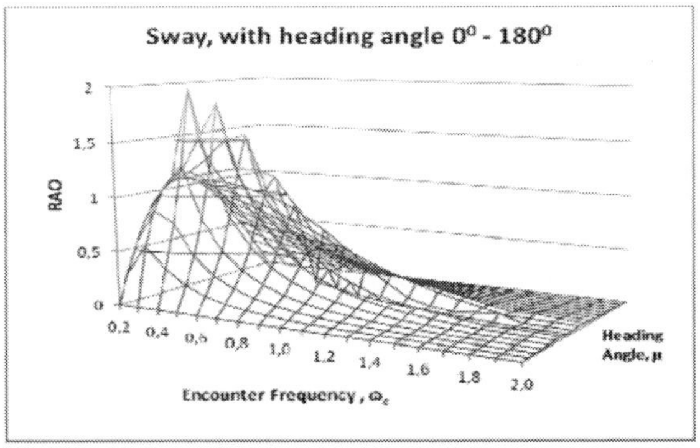

Figure 4 RAO of Sway motion with heading angle 0^0 - 180^0

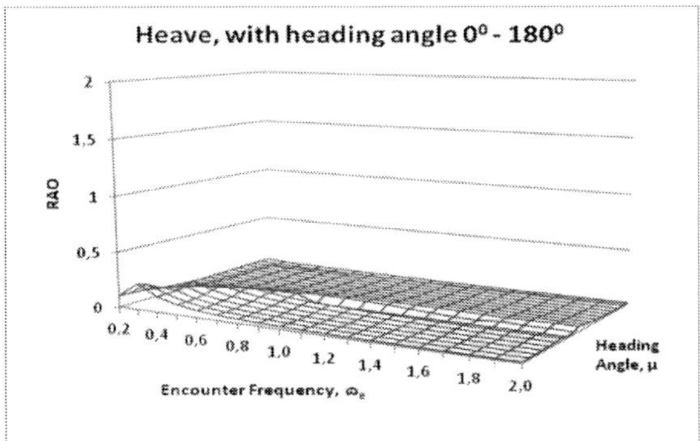

Figure 5 RAO of Heave motion with heading angle 0^0 - 180^0

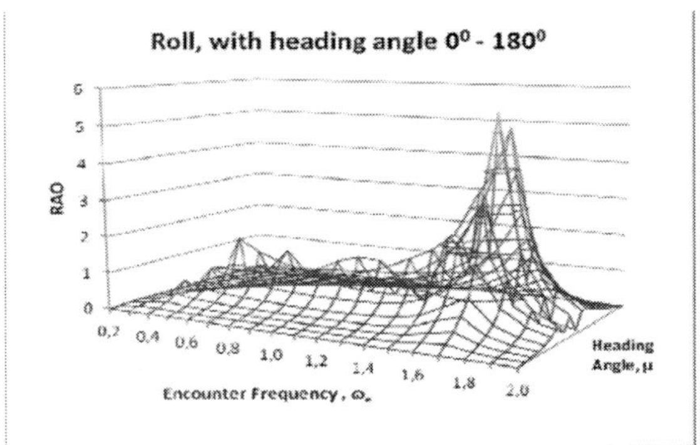

Figure 6 RAO of Roll motion with heading angle 0^0 - 180^0

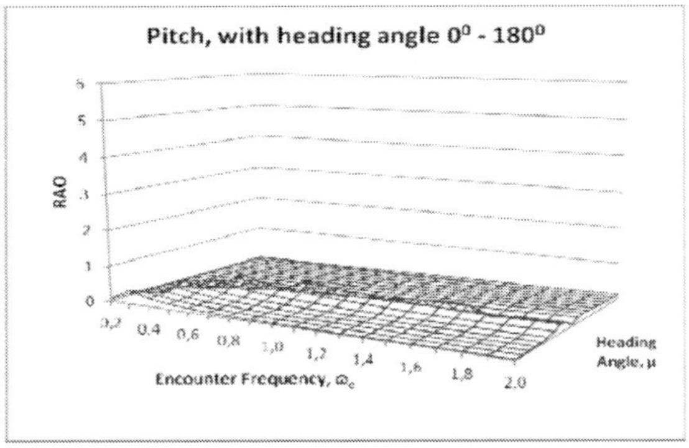

Figure 7 RAO of Pitch motion with heading angle 0^0 - 180^0

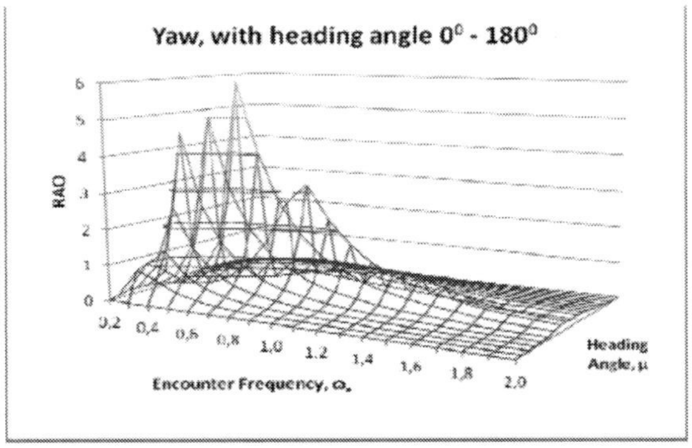

Figure 8 RAO of Yaw motion with heading angle 0^0 - 180^0

3. ENVIRONMENTAL CONDITION

The Trimaran ferry ship was designed for a purpose of transportation on the route of Kalianget, Madura Island and Kangean Island as presented on Figure 9.

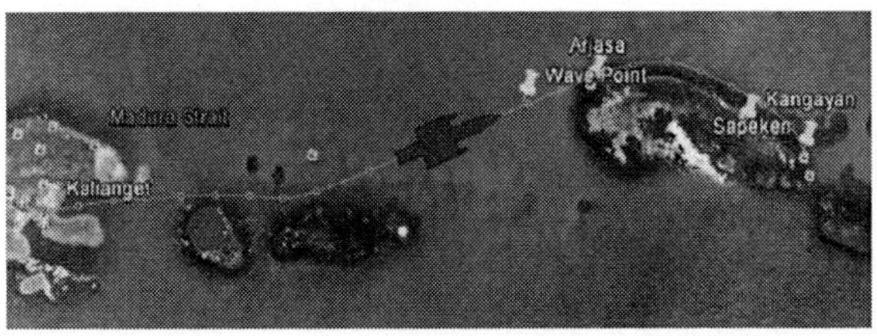

Figure 9 Route of ship voyage from Kalianget to Kangean

In this case, the environmental condition was considered for 1 years time period which was pointed on the operational route of the Ferry ship between Kalianget and Kangean, see Figure 9. The environmental data was described into the snap shot data for each 1 hour with totalling number of 8760 cases. The rule direction of the wave was determined as zero when the wave coming from the north, and it increased clockwise, see Figure 10.

The occurrence probability of wave height had above 3 m was about 0.16 percent of one-year wave height, and the remaining about 1.92 percent of wave height had from 2 m up to 3 m, and 97.92 percent under 2 m. The direction of wave propagation was predominantly coming from the north-west and east. In the case of the ship operate on the route of Sumenep – Kangean, the heading angle of ship was coming dominantly from the direction of quartering and broaching.

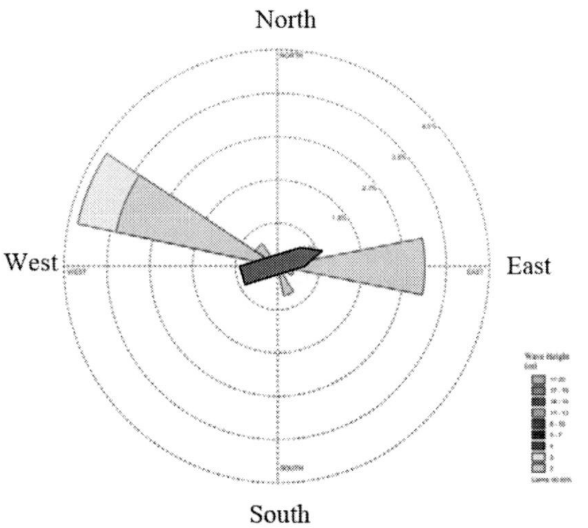

Figure 10 Height and direction of wave on Sumenep – Kangean

For simplicity analysis, the significant wave height was considered. The highest wave height was selected from a week data during one year period of time, so that there were number of 52 snap shot data used for motion analysis. The irregular long crested waves were modelled using the JONSWAP spectrum. The generalized five-parameter JONSWAP formula can be written as:

$$S_{\zeta}(f) = \alpha H_s^{\,2} \frac{f^{-5}}{f_p^{-4}} \exp\left[-\frac{5}{4}\left(\frac{f}{f_p}\right)^{-4}\right] \gamma^{\exp\left[-\frac{(f-f_p)^2}{2\sigma^2 f_p^2}\right]}$$

Where $\alpha = \dfrac{0.0624}{0.230 + 0.0336\,\gamma - 0.185\,(1.9+\gamma)^{-1}}$

$\sigma_A = 0.07$; for $f \leq f_p$

$\sigma_B = 0.09$; for $f > f_p$,

$S_{\zeta}(f)$ is the energy spectrum over frequency f (H_z). The formulation has five free parameters, being significant wave height H_s (m), peak period T_p (sec), peak frequency f_p (1/sec), Phillips parameter α, spectral peakedness parameter γ, low frequency width parameter σ_A, and high frequency width parameter σ_B. The JONSWAP spectral parameterization equations can be defined from the snapshot data available to satisfy above equation.

The encountered wave spectrum of Sumenep – Kangean route were presented in Figure 11 for the 52 weeks snap shot data of a year. The Figure 11 shows the highest amplitude of motion response was on the month of February at frequency about 1 rad/sec.

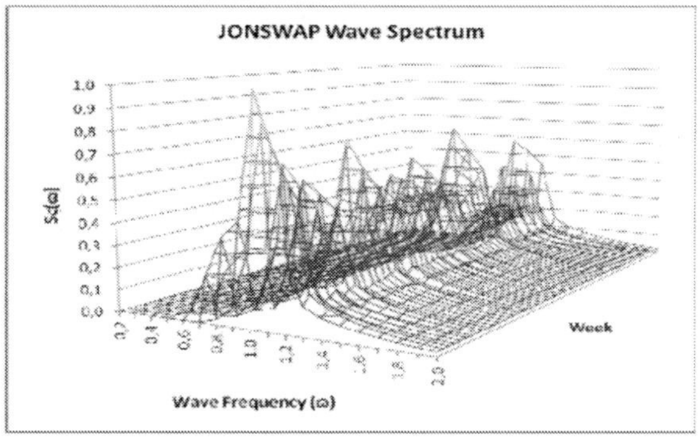

Figure 11 Wave Spectrum on the Sumenep – Kangean route

4. SHIP MOTION IN IRREGULAR WAVE

The Trimaran Ferry Ship's motion responses in multi-directional seaway were calculated by the multiplication of the Response Amplitude Operator (RAO) and multidirectional encountered wave spectra obtained from the environmental data. The equation of standard deviations and period of each motion were obtained as follow,

Standard deviation, $\sigma = \sqrt{m_0}$

Period, $T = 2\pi \dfrac{m_0}{m_1}$

Significant amplitude response, $A_{1/3} = 2\sqrt{m_0}$

where $m_0 = \int_0^{\infty} S(\omega_e) RAO\, d\omega_e$

$$m_1 = \int_0^{\infty} \omega S(\omega_e) RAO\, d\omega_e$$

The significant amplitude response of motions in 52 cases of environmental condition at the route of Sumenep – Kangean were shown in Figure 12. The highest responses of motions were shown at Sway with an amplitude up to 0.3 meter, and the Roll motion with amplitude close to 1.5 degree. And the lowest responses of translation and rotation motions were shown at Surge and Pitch with the amplitude value less than 0.05 m or 0.05 degree respectively.

The highest significant response of Yaw motion was on a particular week of 9th (on March), and the heave response was almost stable for the whole weeks. The significant responses of motion were having variation value for the entire week as same as the condition of wave height had variation for a year.

The motion responses for 52 cases were scattered in terms of period and standard deviations as shown in Figure 13.

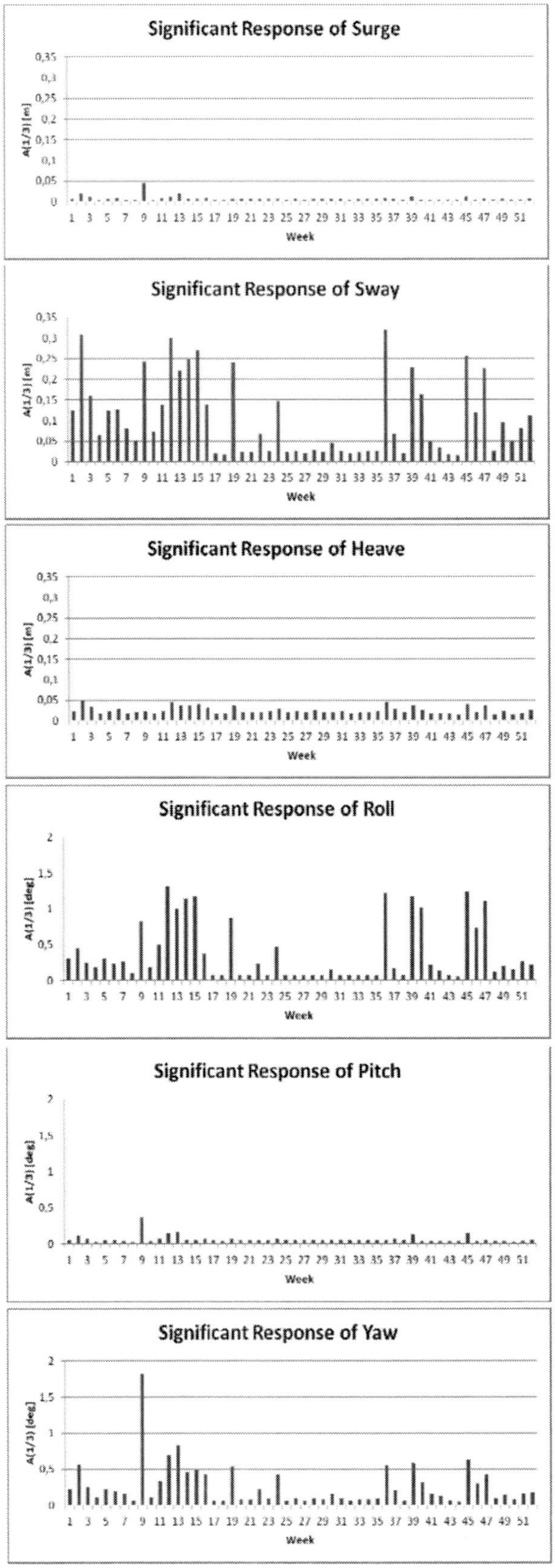

Figure 12 Significant Response of Motions on Sumenep – Kangean route

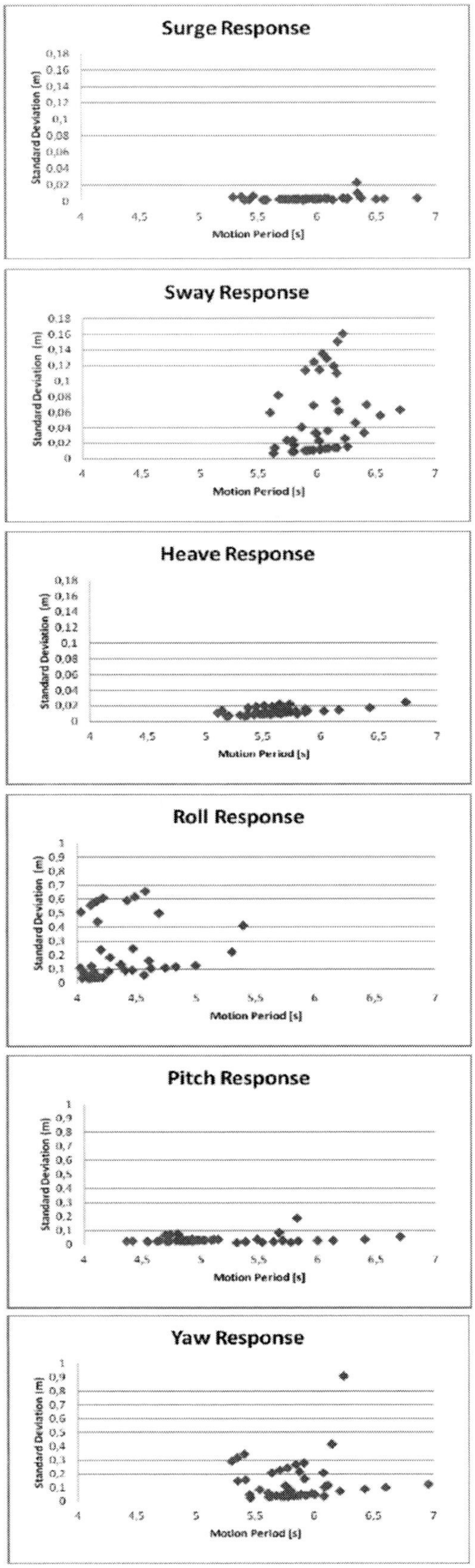

Figure 13 Scatter of standard deviation and period

Figure 13 had shown the response of Surge, Sway, Heave and Yaw motion scatter from period of 5 second through 6.5 second, and the roll motion was scattered on period of 4 second to 5.5 second. The highest response of Surge, Sway and Yaw was on period of about 6.4 second, and the Roll motion had a highest response while the period of about 4.5 second.

The cumulative probability for response motions to exceed certain values in an environmental condition were derived, and presented on Figure 14 through Figure 19. For the design consideration, an extreme value for a desired probability of being exceeded was examined.

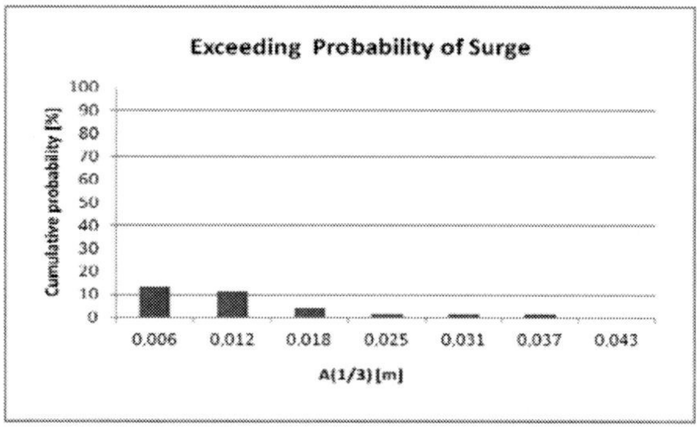

Figure 14 Cumulative exceeding probability of Surge

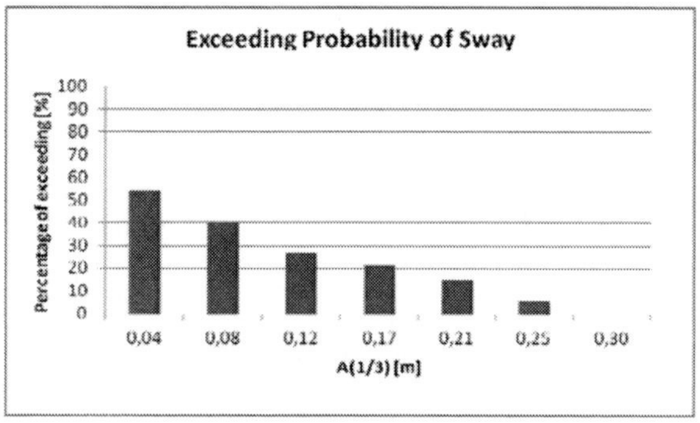

Figure 15 Cumulative exceeding probability of Sway

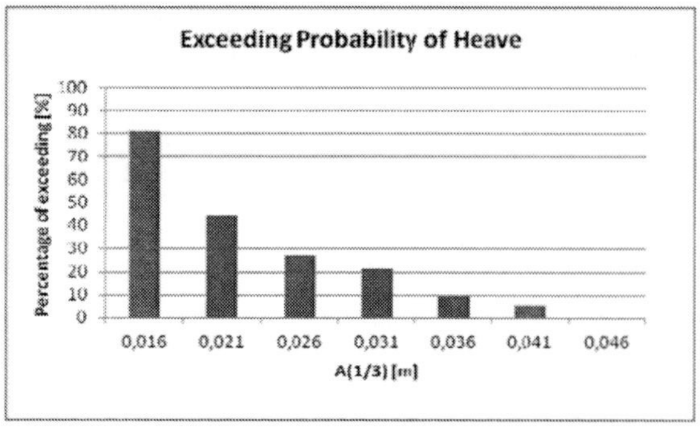

Figure 16 Cumulative exceeding probability of Heave

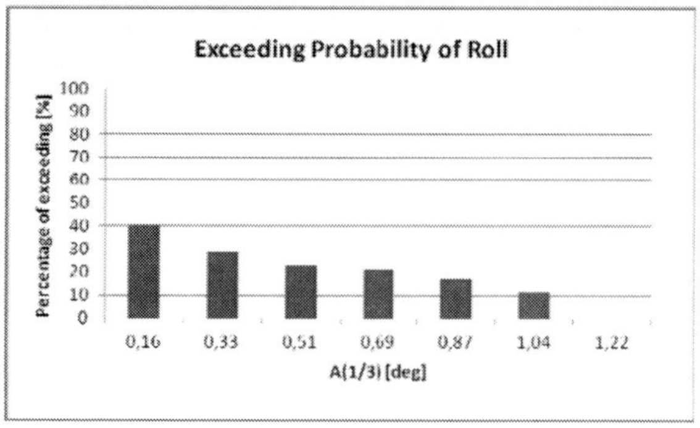

Figure 17 Cumulative exceeding probability of Roll

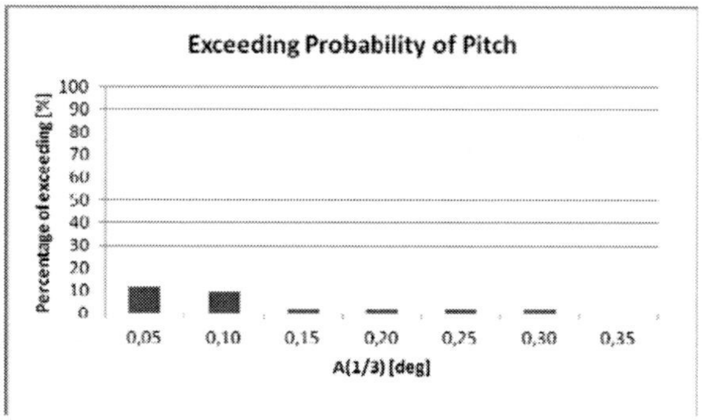

Figure 18 Cumulative exceeding probability of Pitch

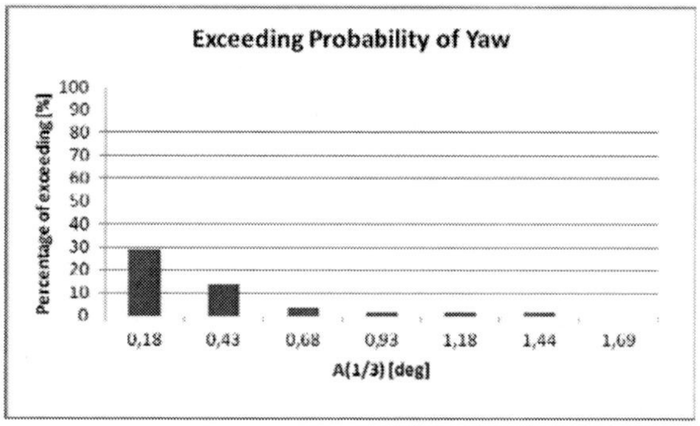

Figure 19 Cumulative exceeding probability of Yaw

The highest response of all motion presented on Figure 14 through Figure 19 were occurred probability less than 5 percent, and an exception of Roll motion occurred up to 11.54 percent. The maximum significant response of roll motion was about 1.04 degree which was occurred 11 percent of a year period. The maximum response of Yaw motion which was presented on Figure 19 had value of 1.44 degree, and it only occurred 1.92 percent for a year.

In over all the response motions of Trimaran ferry ship which was examined on the environmental wave condition of Sumenep – Kangean route, were slightly small, and ship can be operated for the whole year.

5. CONCLUSIONS

Seakeeping analysis was performed due to the design of trimaran ferry ship that was operated on the east java sea, from Sumenep Madura Island travel to the Kangean Island directly. A new approach was proposed in order to examine the seakeeping performance of vessel on the short crested seaway for a year period of time. The high response of motion in regular wave was shown in Sway and Yaw at low frequency, and Roll at high frequency. Based on the environmental condition, the heading relative of ship due to wave direction dominantly was coming from quartering and broaching. The examination results shown the trimaran ferry ship had a good seakeeping performance, and it was allowed operating on the route of Sumenep – Kangean for the whole year.

6. ACKNOWLEDGMENTS

The author would like to express thanks to the Incentive Research Award 2011 from the Ministry of Research and Technology of Indonesia for financial support.

7. REFERENCES

1. Davis, M.R., and Holloway, D.S., "Motion and Passenger Discomfort on High Speed Catamarans in Oblique Seas," International Shipbuilding Progress, 50, no. 4, pp. 333-370, 2003.

2. Faltinsen, O.M., "Hydrodynamics of High - Speed Marine Vehicles," Cambridge University Press 2005.

3. Lyakhovitsky, A, "Shallow Water and Supercritical Ships". Backbone Publishing Company, 2007.

4. Utama, IKAP; Murdjanto; Sulisetyono, A; and Jamaludin, A, "Pengembangan Kapal Hemat Energi Dengan Konfigurasi Kapal Berbadan Banyak", Seminar on Application and Research in Industrial Technology, UGM, Yogyakarta, 27 Agustus 2008.

5. Sulisetyono, A; Nasirudin, A; and Hermawan, Y. A, " Perancangan Kapal Penumpang Barang Tipe Trimaran untuk Pelayaran Antar Pulau," Seminar Nasional Teori dan Aplikasi Kelautan (SENTA), ITS Surabaya, December 2011.

6. Sulisetyono, A, "An Analytical Approach of the Free Surface Green's Function for Radiation Problem", International Conference on Ship and Offshore Technology (ICSOT), ITS Surabaya, 2010.

7. Sulisetyono, A, "Single Body Motion Analysis", Project report of FLNG Mooring and Offloading Study, LPPM ITS, 2009.

8. AUTHOR'S BIOGRAPHY

Aries Sulisetyono holds the current position as Head of The Centre for Marine Science and Technology Studies, Institut Teknologi Sepuluh Nopember (ITS), Surabaya, Indonesia. He is responsible for teaching in the subject of Ship Dynamics at the department of Naval Architecture and Shipbuilding Engineering. In 2005 he obtained a PhD degree in Naval architecture and Marine Engineering from Dalhousie University Canada, on the topic of wave-body interaction problem. Apart from performing a PhD-research, his tasks included teaching assistance of Fluid Mechanics course, serving as assistance towing tank laboratory, and researches at CMVDR (Centre for Marine Vessel Development and Research) Dalhousie University.

AUTHORS' CONTACT DETAILS

Andi Jamaluddin
Indonesian Hydrodynamic Laboratory (IHL) Surabaya, Indonesia
andi99@engineer.com

Ferry Manuhutu
Universiti Malaysia Terengganu, Malaysia
ferdomanu@yahoo.com

Peter Sanders
BMT Design & Technology, Melbourne, Australia
psanders@bmtdt.com.au

Kazuhiko Shiotani
JFE Steel Corporation, Japan
k-shiotani@jfe-steel.co.jp

Ketut Utama
ITS, Surabaya, Indonesia
kutama@na.its.ac.id

Arief Suroso
Dept of Ocean Engineering-ITS, Indonesia
arief_suroso@yahoo.com

Wolter R Hetharia
Université de Liége, Belgium
hethariawr@yahoo.com

Aries Sulisetyono
Institut Teknologi Sepuluh Nopember (ITS) Indonesia
sulisea@na.its.ac.id

Clive Evans
Lightning Naval Architecture, Australia
navlight@lna.com.au

Daeng Paroka
Hasanuddin University, Indonesia
d_paroka@yahoo.com

Bastien Abeil
Maritime Research Institute Netherlands, The Netherlands
b.abeil@marin.nl

Muhammad Alham Djabbar
Hasanuddin University, Indonesia
alham.djabbar@yahoo.com

Sunaryo
Department of Mechanical Engineering, University of Indonesia
naryo@eng.ui.ac.id

Daniel M Rosyid
Department of Ocean Engineering, ITS Surabaya, Indonesia
dmrosyid@gmail.com

Johannes M Herman
Marine Engineering, Pattimura University
herman_tsp_unpatti@yahoo.com

Eka Taniputra
Orela Shipyard, Indonesia
eka@orela.co.id

Richard Birmingham
Newcastle University, UK
r.w.birmingham@newcastle.ac.uk